The Home Workshop SPY

Spookware
for the
Serious Hobbyist

Nick Chiaroscuro

PALADIN PRESS • BOULDER, COLORADO

The Home Workshop Spy:
Spookware for the Serious Hobbyist
by Nick Chiaroscuro

ISBN 0-87364-922-2
Printed in the United States of America

Published by Paladin Press, a division of
Paladin Enterprises, Inc., P.O. Box 1307,
Boulder, Colorado 80306, USA.
(303) 443-7250

Direct inquiries and/or orders to the above address.

PALADIN, PALADIN PRESS, and the "horse head" design
are trademarks belonging to Paladin Enterprises and
registered in United States Patent and Trademark Office.

CONTENTS

WARNING

This book does not intend to imply, by its lurid title or its cavalier take on the matter, that electronic surveillance is noble, legal, or safe. It does not seek to promote bugging or to lure anyone into behavior that could violate the rights of others and subject the perpetrator to criminal prosecution. *This book seeks only to inform the reader.* Those who act upon the information contained herein do so at their own risk and entirely on their own responsibility.

The author and the publisher specifically disclaim any and all liability and responsibility to any party for damage or loss caused, or alleged to be caused, directly or indirectly by use, misuse, or inability to use the information contained herein. Those who elect to build and/or use these projects do so at their own risk and entirely on their own responsibility.

PREFACE

README.NOW

This book is not for beginners. It presumes the reader to be a responsible adult, 21 years or older, with:

- working knowledge of electronics and related math
- access to a digital multimeter and an oscilloscope
- skill in the use of hand tools
- soldering skill
- skills needed to fabricate printed circuit boards
- the ability to work with common materials, including plastic and metal
- the capacity to manage all tasks safely
- the sense not to break the law in the process

Those who lack any of these qualifications should not attempt to build or use these projects.

The text uses the term "wire specialist" to mean *a person with a legal right to build, possess, transport, and use the tools of electronic surveillance*. Despite the hyperbole implicit in a casual treatment of bugging, readers are cautioned that the projects work substantially as described. It is a crime for unauthorized personnel to eavesdrop, through any means, on persons without their knowledge and consent. Readers should also note that the mere possession of a device capable of eavesdropping could, in the eyes of some officials, constitute a crime.

The nature of the topic and the prevalence of entrapment make it impossible for the author to respond to queries. Don't waste a stamp.

All prototypes were destroyed prior to the publication of this book. No functional project or accessory remains in the author's possession. All logs, photos, notes, etc. relating to the production of this book, including those that might have resided on a computer system, have been destroyed.

Finally, the reader should understand that some sects of the law enforcement community could regard possession of this book as proof of imminent sin. Money trails and mailing lists relating to its purchase might bear the same taint. The author presents this data for lawful use only, as in the hands of those empowered by law to wrangle forbidden science, or for hobbyists and experimenters who wish to study the concepts involved, or for casual readers who wish to acquaint themselves with surveillance hardware. *Readers are warned not to break the law*. Readers who are unsure about the legality of specific behavior should consult competent legal counsel before undertaking the behavior.

INTRODUCTION

GENERAL CONSIDERATIONS

PARTS

Projects worked as described using the specified parts. In most cases they will work equally well using electrically equivalent parts, but the text has not tested parts other than those listed.

Capacitors' rated working voltage must match that of the circuit. Boards have been laid out to accommodate electrolytic caps having the lowest working voltage higher than the supply. "Ceramic bypass" refers to -30 percent/+80 percent (temperature unstable) types. "Coupling" caps should be 20 percent or better. Frequency-determining caps, as used in active filters and tank circuits, should be temperature stable, 5 percent or better "poly" types (polyethylene, polypropylene, polystyrene, polycarbonate).

All resistors are 5 percent ¼ W or ⅛ W types unless otherwise specified.

Variable inductor pads are sized for the Toko 10.5 mm footprint.

Audio output stages are designed for low-impedance, high-efficiency, sealed mono headphones.

Stuffing and wiring diagrams use a single model of a jack, resembling a plastic ⅛-inch type. For microphone/sensor inputs, the text recommends RCA (phono) type jacks and plugs. Cable used to test mikes and sensors was RG174/U coax.

CIRCUIT BOARDS

All circuit board patterns are shown foil side, 100 percent size. Reference lines are printed with each to allow compensation for distortion that might have occurred in the printing process.

The boards have been laid out to facilitate construction using rub-on pads and tape. Prototypes were built by copying the board layout full-size, then taping this template securely to the copper side of a board pre-cut to the desired size. Holes were drilled in the center of each pad through the board beneath. The template was then removed intact and both sides of the board smoothed with sandpaper and abrasive kitchen cleanser. Etch-resist pads and traces were applied to the drilled board using the copied pattern as a visual guide, then etched in the usual fashion.

GOOF-PROOFING THE POWER BUS

The text assumes the reader to know that accidentally reversing power polarity, as when connecting a battery, can destroy semiconductors. Those who wish to protect expensive chips from accidents can place a rectifier diode (e.g., 1N4004) in series with the appropriate power lead:

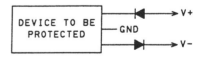

No significant current flows in circuits so protected unless battery polarity is correct. The drawback of this technique is that it drops the power supply by about 0.6

1

V. It best suits circuits with multiple, costly integrated circuits and will not protect incorrectly mounted parts.

PROTOTYPES

Close examination of prototype photos will show that some of them do not follow the board stuffing and wiring diagrams exactly. Several diagrams were altered after construction of the prototype showed that changes would improve the device. Socketed chips are those that happened to be present when photos were taken, not necessarily those recommended in the parts list.

All prototypes were built in anticipation of their being destroyed. This let them eschew such niceties as metal cases, attractive knobs, labeled panels, etc.

OTHER

The wire specialist must consider that these projects' parts profiles could reside in the scanning databases of several intelligence services and that ordering certain combinations of parts might trip an alarm in dark quarters. Discounting the lunacy behind this speculation, but noting the upshots that could accrue if it were true, the wire specialist might choose to adjust his affairs.

Once it nears finished form a project should bear an indelible label that identifies it as an EXPERIMENTAL ELECTRONIC DEVICE—NOT FOR SALE OR TRANSFER. Unlabeled tools have a way of becoming grand jury exhibits.

Having served its purpose, assuming it wasn't sacrificed in the job, a project must be neutralized by more than desocketing chips or unsnapping a battery. Prototype boards were gutted and halved, then discarded.

Any documentation left in the course of building a project should reflect some legitimate hobby, science fair project, or the like.

Open testing cannot be condoned. Exotica has a way of calling attention to itself. Use of confederates is also unwise, observing the wire specialist's axiom that risk of discovery grows as the cube of the number of people who know. Experimenters who effect some familiarity with bugging tools will eventually find themselves mulling over an offer to build/sell/use those tools. The proposal will come, absolutely and positively, from an entrapment snitch.

Wire specialists who build surveillance gear have little to fear, for the term connotes official sanction. Others who dabble in this forbidden realm have pretty well exiled their legal rights to the *Twilight Zone*. It follows that those unprepared to deal with strange developments should not walk on the wild side.

CORNER DIRECTIONAL MICROPHONE

Directional mikes don't have to bc big to be effective. They do have to deliver gain based on physical principles rather than simple exclusion of off-axis sound. The corner directional microphone (CDM) uses the boundary effect to achieve 18 dB of contoured, directional gain.

An ancient but still formidable LM381 serves as preamp. Old hands will see in the schematic a twist on the widely hailed Carter-era circuit whose fantastic potential never seemed to surface in any sort of whole and usable tool.

All strong sound amps need dynamic control. The CDM uses a true limiter adapted from a concept shown in Signetics' *Linear Data Manual Volume 1: Communications* (p. 4-344). It attacks in ~1 msec, decays in ~100 msec. It cycles with each peak. Switching transients can become intrusive if threshold is set so low as to moderate constant tones.

CDM PARTS LIST

Capacitors
C1, 6, 20, 24 220 μF aluminum
   electrolytic
C2, 4, 5, 9, 12, 13, 18 10 μF aluminum
   electrolytic
C3 2.2 μF aluminum electrolytic
C7, 11, 14 100 pF ceramic bypass
C8, 17, 19, 22 0.1 μF coupling
C10, C21 0.001 μF ceramic bypass
C15, 16 150 pF ceramic bypass
C23 0.1 μF ceramic bypass

Resistors
R1 4.7 K
R2 200
R3 36
R4 36 K
R5 5 K multiturn miniature trimpot
R6 47 K
R7, 21 39 K
R8, 14, 15, 19 100 K
R9, R10 10 K
R11 2 K
R12 10 K audio-taper pot
R13 6.8 K
R16, 17 100
R18 1 M
R20 470
R22 10 K audio-taper pot w/switch
R23 10

Semiconductors
Q1 2N3906 PNP transistor
U1 LM381 low-noise dual preamplifier
U2 NE570 or NE571 dual compandor
U3 LM393 dual comparator
U4 LM78L05 5 V positive regulator
U5 LM386 audio power amplifier

Miscellaneous
M1 electret condenser microphone
   (Panasonic WM52BM [preferred] or
   Radio Shack 270-090)
S1 SPST switch (part of R22)
15-volt power source, hardware,
   enclosure, shielded microphone
   cable, printed circuit board, etc.

CDM Schematic

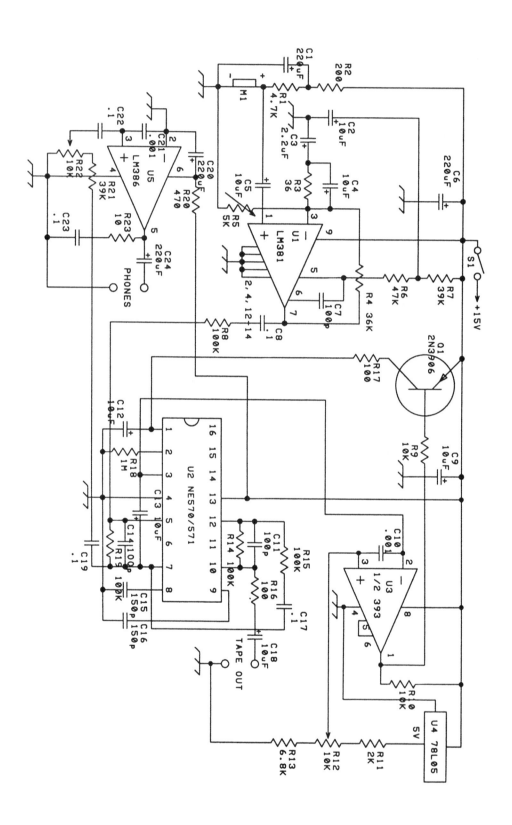

CIRCUIT FUNCTION

Electret condenser microphone M1 gets its bias through R1. M1 output couples through C5 to noninverting input of preamp U1 configured in single-ended mode by grounding unused input pin 2. Minimum noise occurs with 381 input transistor collector current at ~170 μa, set externally by R6 in series with R7. C2 decouples this path. Input pins 12-14 of U1's unused channel are grounded.

Gain of U1 is set by the ratio of R4 to the net impedance of the network formed by R3, R5, C3, and C4. This varies with frequency, giving a sharp treble emphasis. R5 also sets DC offset of output pin 7. C7 rolls off U1's response above 10 kHz.

Output of U1 couples through C8 and R8 to input of one channel of U2, configured for unity gain in its resting state by the ratio of R19/R8. C14 limits the high-frequency response of U2's internal op amp. R19 also sets the DC bias point of U2 output pin 7 close to ½ V+, preserving maximum dynamic range.

U2 runs on an internal bias of ~1.8 volts. This bias is present at pin 3. C13 impresses U2's audio output on the resting bias, tied to input pin 2 of dual comparator U3, shunted by C10. A resistive divider formed by R11, R13, and pot R12, on supply locally regulated by U4, allows U3 trip threshold to be altered by R12. R10 is the comparator pull-up resistor; comparator output couples through R9 to base of Q1.

When the comparator output is high, Q1 is turned off. In this state U2's control channel passes audio at unity gain. When comparator output goes low, Q1 turns on. This causes U2's control channel to cut gain. Attack is very fast, on the order of a millisecond, set by the time constant of R17 and C12. Decay is determined by the time constant of C12 and a 10K resistance internal to U2. R18 leaks current to keep C12 partly charged. Input pins 5 and 6 of

U3's unused comparator are shorted.

Signal leaving the limiter channel of U2 passes through C17 and R15 to the op amp in U2's second channel, which serves as a tape output buffer. Its gain is set at unity by the ratio of R14/R15. C11 limits the high-frequency response of the buffer. Signal couples through R16 and C18 to tape output. C15 and C16 decouple U2's internal supply.

Audio also couples from U2 pin 7 through C19 to resistive divider formed by R21 and volume control R22. This divides out some of the 26 dB voltage gain built into U5, leaving U1 the gain-determining element. Signal couples through C22 to noninverting input of U5. C21 shunts RF at U5 input. R20 and C20 decouple U5 from the supply; R20 also lowers the 15 V supply to about 12 V at U5, allowing 386N-1, -2, and -3 versions to run safely. R23 and C23

CDM Circuit Board

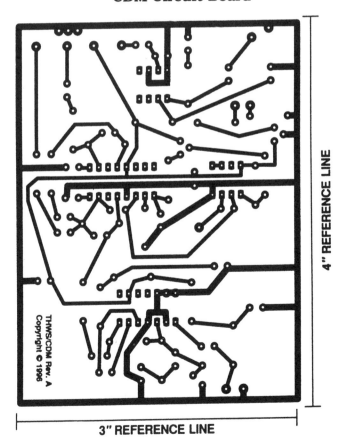

THWS/CDM Rev. A
Copyright © 1996

4" REFERENCE LINE

3" REFERENCE LINE

CDM Stuffing and Wiring Diagram

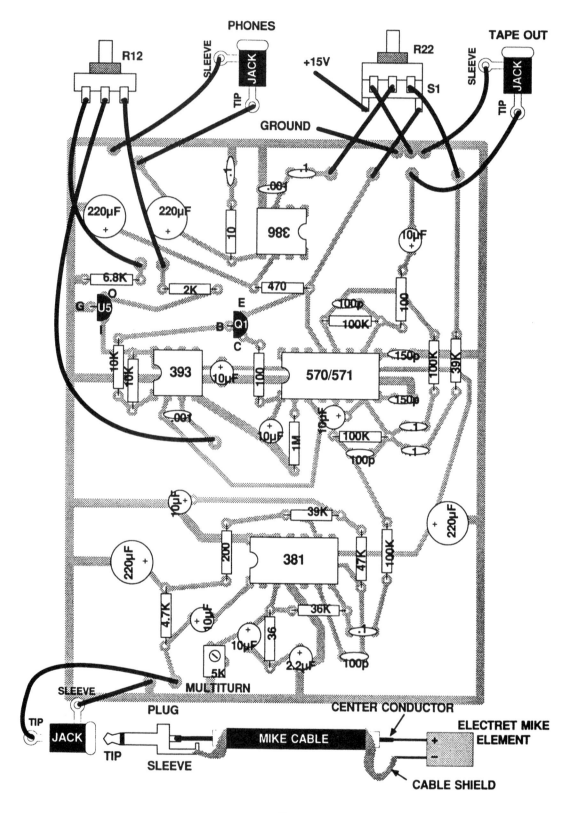

CDM Amplifier Prototype

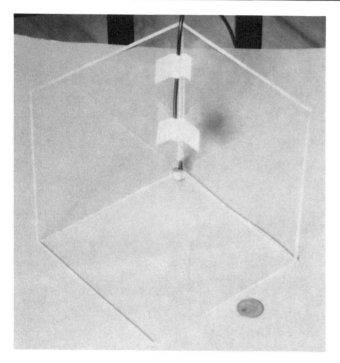

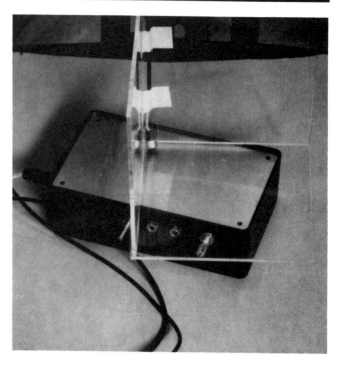

form the standard snubber. Audio couples through C24 to headphone output jack.

R2, C1, C6, and C9 decouple the main supply.

DETAILS

The mike element mounts *face-down* in the corner formed by the intersection of three boundaries. Space between mike and boundary should not exceed 1/16 inch. The thickness of a dime will do. Secure the element with glue for permanent use, or secure just the cable with masking tape for quick disassembly.

Construction details are apparent from inspection: an open corner with 5-inch square sides intersecting at right angles. The prototype used scraps of 1/8-inch acrylic, but any thin, rigid material will work. Joints should be airtight. Sides as small as 3 inches will work; sides as big as a foot improve rejection of ambient noise.

Apply power, trim R5 to give 1/2 V+ at pin 7 of U1.

Before donning the phones it's a good idea to set limiter threshold at minimum (R12 full CCW). Adjust limit threshold for existing conditions.

Raw capsule output of this configuration is about eight times that of the same mike element mounted in a "shotgun" directional mike. The preamp is contoured for greatest gain in the speech band (300 to 6000 Hz). Coupled to EQ supplied by the boundary, the system needs no highpass filter despite condenser mikes' great bass response.

Like any serious directional piece, the CDM should not be hand-held or used in a noisy milieu. It works best mounted on a tripod on a cold, still night.

So here it is: tweaked for speech, in full spook trim, rebiased for battery power, a mere 20 years too late, the ancient classic now born of flesh, the one—the *only*—LM381 "ultra-low-noise preamplifier."

PROJECT NO. 2

SONARBUG

Strides in RF and IR debugging drove wire specialists to media that escape detection in standard sweeps. One such medium is ultrasonic audio. The sonarbug (SNB) frequency modulates a 40 kHz audio carrier.

CIRCUIT FUNCTION

Microphone M1 gets its bias through R1. M1 output couples through C1 to noninverting input of U1, biased through R5 to ½ V+ produced by voltage divider R1-R2. U1 is configured as a noninverting amp with frequency-dependent gain, set by the ratio of feedback resistor R6 to the network formed by R4, C4, and C5. C3 rolls off the response above 4 kHz. D1 and D2 clip U1 output at about 0.7 Vp-p. U1 output couples through lowpass network R7-C7 to C8, then to pin 5 of U2, a CMOS 555 timer configured as an astable multivibrator whose free-running frequency is set close to 40 kHz by R9, R10, and C10, trimmed by pot R8. U2 runs on a supply locally regulated at 5 V by U3. U2 drives the ultrasonic transducer directly.

C1, C2, and C9 decouple the supply.

DETAILS

The SNB requires a supply of at least 6 V but not more than 15 V (10 V for a MAX410).

M1 and the ultrasonic transducer can mount off-board, using high-quality cable preferably shorter than 3 feet.

Tune-up, transducers, and other

SNB PARTS LIST

Capacitors
C1, 2, 9 220 μF aluminum electrolytic
C3, 8 0.1 μF coupling
C4 10 μF aluminum electrolytic
C5 2.2 μF aluminum electrolytic
C6 0.0033 μF 10 percent or better
C7 0.01 μF 10 percent or better
C10 0.0015 μF 5 percent or better

Resistors
R1, 2, 6, 10 10 K
R3, 9 2.2 K
R4 100
R5 22 K
R7 4.7 K
R8 10 K single-turn ¼-inch trimpot

Semiconductors
D1, 2 1N34A
U1 MAX410 or TL071 or LT1097 op amp
U2 LMC555 CMOS timer
U3 LM78L05 5 V positive regulator

Miscellaneous
J1 jumper
M1 electret condenser mike (Radio Shack p/n 270-090 or equivalent)
Ultrasonic transducer (Panasonic p/n EFR-TUB40K5 or equivalent)
9 V battery, printed circuit board, solder, wire, etc.

particulars are discussed in conjunction with the sonarbug receiver.

SNB Schematic

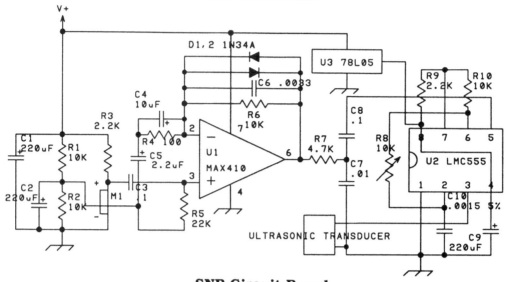

SNB Circuit Board

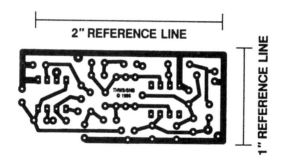

2" REFERENCE LINE

1" REFERENCE LINE

SNB Stuffing and Wiring Diagram

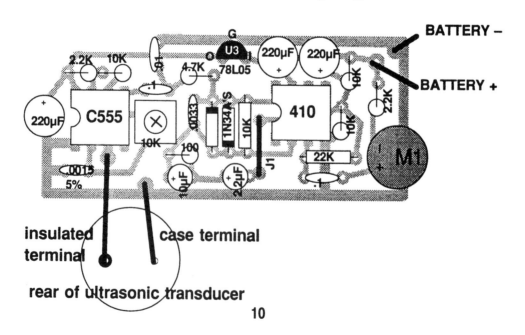

BATTERY –

BATTERY +

insulated terminal   case terminal

rear of ultrasonic transducer

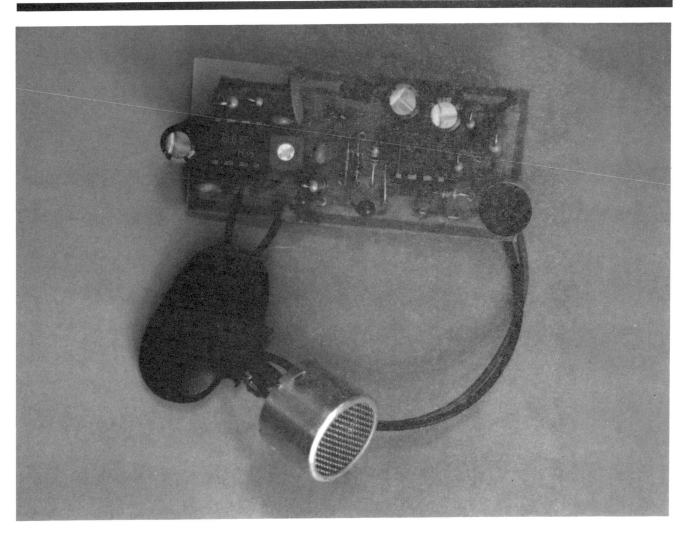

SNB Prototype

PROJECT NO. 3

SONARBUG RECEIVER

Companion to the sonarbug.

SYSTEM THEORY

The transducers are co-resonant near 40 kHz. If the carrier is tuned slightly away from resonance, frequency modulation of the carrier by audio generates amplitude modulation at the sending transducer and the receiving transducer. The pair acts as a slope detector. The receiver recovers audio by rectifying and lowpass filtering the carrier.

CIRCUIT FUNCTION

Ultrasonic transducer couples to the base of Q1, configured as a common-emitter amplifier by R1. S2 selects collector bias through R2 or tank L1-C2. Q1 output couples through C3 to base of Q2 configured as a common-emitter amp by R4, R5, and R7. C23 rolls off response above 200 kHz. Q2 output couples through C5 and R8 to op amp U3, configured as a precision halfwave rectifier by D1, D2, and R9.

SBR PARTS LIST

Capacitors
C1, 4, 15, 16, 21, 22 220 μF aluminum electrolytic
C2 0.01 μF, 5 percent or better poly
C3, 20 10 μF aluminum electrolytic
C5, 6, 13, 18 0.1 μF coupling

C7, 11, 12 0.01 μF, 10 percent or better
C10 0.1 μF, 10 percent or better
C14 0.001 μF ceramic bypass
C17 0.1 μF ceramic bypass
C19, 23 150 pF ceramic bypass

Resistors
R1, 4 1 M
R2, 14, 20 2.2 K
R3, 7, 18, 22 100
R5, 11 4.7 K
R10, 12, 13 47 K
R15, 16 22 K
R17 10 K audio-taper pot
R6, 19 10
R21 220 K
R23, 24 10 K

Semiconductors
D1, 2, 3, 4 1N914 or 1N4148
Q1, 2 2N3904 NPN transistor
U1 LF444 quad low-power op amp
U2 LM386 audio power driver
U3 MAX439 low power op amp

Miscellaneous
J1, 2 jumper
L1 2.2 mH variable inductor (Digi-Key p/n TK1707)
S1 SPST switch (part of R17)
S2 SPDT switch
Ultrasonic transducer (Panasonic p/n EFR-RUB40K5 or equivalent)
9 V battery, printed circuit board, solder, wire, etc.

U3 output (taken at resistor/diode juncture) couples through C6 to U1-a, configured as a quasi-18 dB/octave lowpass filter by R10-13 and C7-9. Cutoff frequency is just above 3 kHz. U1-a output couples to U1-b, configured as a quasi-18 dB/octave highpass filter by C10-12 and R14-16. Cutoff is around 700 Hz.

U1-b output couples through C18 to U1-c, configured by R20 and R21 as an inverting buffer with gain of 40 dB. C19 limits high-frequency response; diodes D3 and D4 clip the output of the buffer at about 1.2 Vp-p. U1-c output couples

through R22 and C20 to tape output jack.

U1-b output also couples through C13 to volume control pot R17, whose wiper couples to the noninverting input of U2. C14 shunts RF at U2 input. R19 and C17 form the standard snubber. Audio couples through C16 to headphone output jack.

U1-d is configured as a DC voltage follower whose noninverting input is biased at ½ V+ by divider R23-R24. U1-d output serves as a stable bias reference for U3 and U1-a/b/c.

C1, C4, C15, C21, C22, R3, R6, and R18 decouple the supply.

SBR Schematic

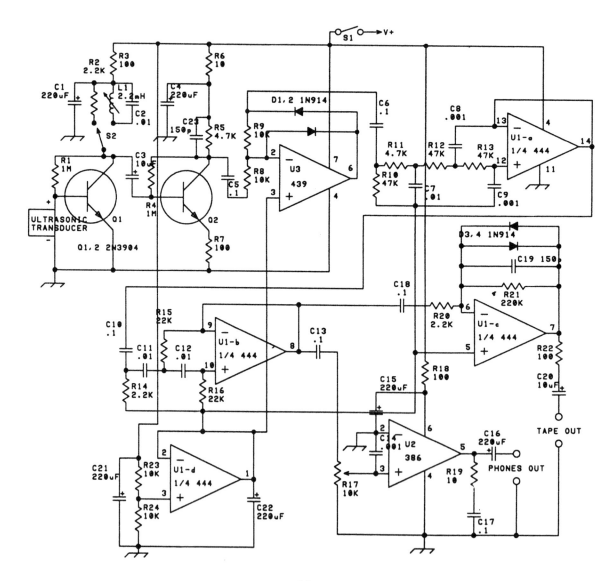

DETAILS

Ultrasonic input jack and bias switch S2 should mount as close as possible to the board. Although the prototype was housed in a plastic case, a metal case is recommended. Supply is 6 to 12 V.

Alignment requires the SNB. Set the sonarbug receiver (SBR) input bias switch to "resistor." Power up both pieces, don headphones connected to SBR. Audio transmitted by SNB should be heard. The pair working in proximity usually overloads the SBR, resulting in distorted audio. Separate the units enough to avoid overload. Trim SNB pot R8 for best audio.

Switch SBR input bias to "tank," tune L1 for best audio.

Fine tuning for range and audio quality can be accomplished solo but is facilitated by a helper and should be performed out of doors over an unobstructed path.

Transmitter and receiver transducers should face each other and be held at least 3 feet off the ground. Switch SBR input bias to "resistor" (the SBR mutes for a second or so when bias is switched from "tank" to "resistor") and retune SNB pot R8 to optimize range and audio quality. Switch SBR to "tank" and retune L1 for range and audio quality. The system requires a trade-off: maximum range coincides with muddy sound. By carefully tuning SNB pot R8 the user will find the carrier that best reconciles intelligibility and range. The SNB regulates the C555 supply to prevent battery aging from shifting the carrier.

Line-of-sight range using raw transducers varies 75 to 200 feet. This can be increased several times by a reflective sonic antenna at the receiver. Great results have been had using smooth plastic bowls 4 to 6 inches, available for less than $2 at the nearest supermarket.

SBR Circuit Board

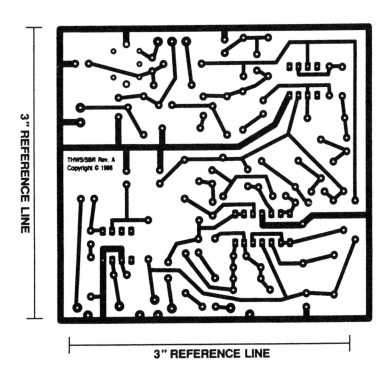

3" REFERENCE LINE

3" REFERENCE LINE

Range possible using an 18-inch parabolic reflector . . . well, the wire specialist can calculate this himself.

With appropriate changes in frequency-determining components the SNB/SBR system will work at frequencies other than 40 kHz and with transducers other than the specified pair.

SBR Stuffing and Wiring Diagram

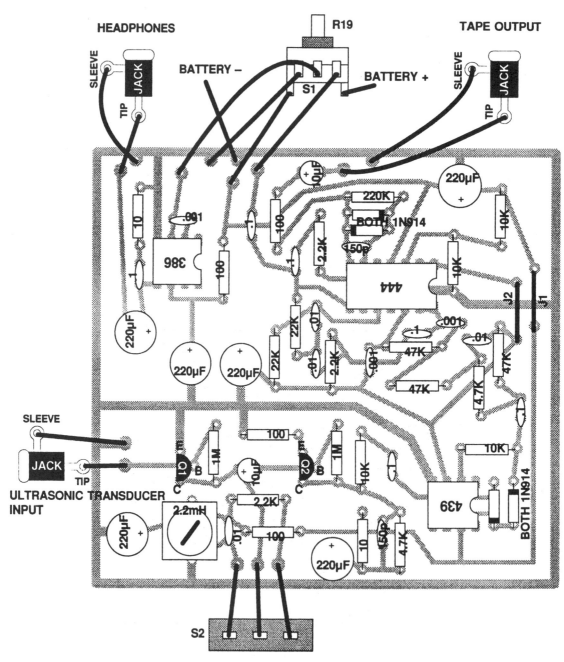

TANK ← INPUT BIAS SELECT → RESISTOR

SBR Prototype

PROJECT NO. 4

HARDWIRE SENDERBUG

Whenever hardwire becomes the mode of choice, as it does a lot more often than the open literature implies, the ultra-versatile hardwire senderbug (HWS) can meet the most demanding application.

CIRCUIT FUNCTION

Microphone M1 gets its bias through R1. M1 output couples through C1 to the base of Q1, configured as a common-emitter amplifier by R2, R3, and R4. Q1 output couples through C4 to base of Q2, configured as a common-emitter amplifier by R5 and R7, with Q2 collector load consisting of the 10 K winding of T1, shunted by R6. C6 takes the edge off the treble emphasis produced by C3 and C5. The 600-ohm winding of T1, loaded by R8, drives the transmission line.

C2, C7, and R9 decouple the supply.

DETAILS

The board is laid out for ⅛ W resistors and miniature caps. Device will function over the range 1.5 to 15 V. From 1.5 to 10 V, output rises with supply voltage. The prototype (built with 3565s) drew 260 ua at 1.5 V; 440 ua at 3 V; 680 ua at 5 V; 980 ua at 7.5 V; 1.3 ma at 10 V; 2.2 ma at 15 V.

A 1,000 mah 3 V lithium coin cell (e.g., Panasonic CR2477-1HF) should run it for more than three months; a 5,000 mah "C"-size lithium cell should power it for longer than a year.

Further details are discussed in conjunction with the hardwire receiver.

HWS PARTS LIST

Capacitors
C1, 4 0.1 μF coupling
C2, 7 10 μF miniature aluminum or tantalum
C3, 5 1 μF tantalum
C6 33 pF ceramic

Resistors
R1 4.7 K
R2, 5 1 M
R3, 4, 6, 7 10 K
R8 1.2 K
R9 200

Semiconductors
Q1, 2 2N3565 or 2N3904 NPN transistor

Miscellaneous
M1 Radio Shack 270-090 electret condenser microphone
T1 600:10 K transformer (Mouser p/n 42TL019 or equivalent)
Power supply, printed circuit board, solder, wire, etc.

HWS Schematic

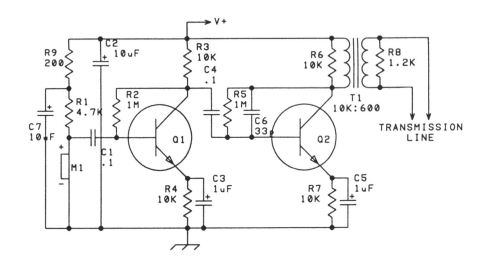

HWS Circuit Board

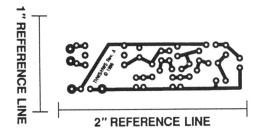

HWS Stuffing and Wiring Diagram

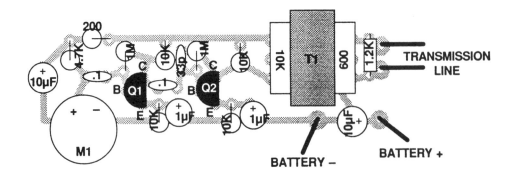

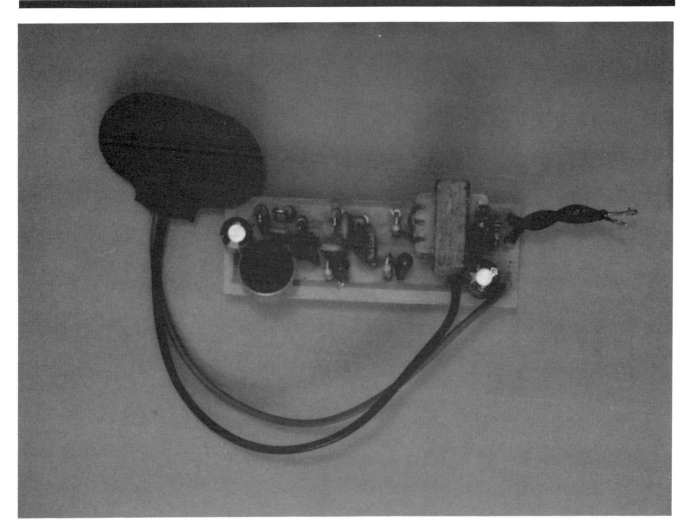

HWS Prototype

HARDWIRE RECEIVER

Companion to the hardwire senderbug.

SYSTEM THEORY

The HWS uses a transformer to let a weakly biased transistor drive a low-impedance transmission line. A transformer steps the voltage down by ~12 dB. The hardwire receiver (HWR) uses the same transformer in reverse, recovering most of the step-down loss. Balanced transmission lets the transformer reject common-mode noise picked up along the way.

CIRCUIT FUNCTION

Balanced transmission line couples directly to 600-ohm winding of T1, loaded by R1; 10 K winding of T1 couples through C3 to input of U1, configured as an inverting amp with variable gain, determined by the ratio of R5 to the net impedance of the transformer winding loaded by R2. D1 and D2 limit U1 input voltage to ~1.2 Vp-p. D3 and D4 limit U1 output to ~1.2 Vp-p. C4 rolls off high-frequency response above a point dependent on setting of R3. Noninverting input of U1 is biased at ½ V+ by divider R3-R4.

HWR PARTS LIST

Capacitors
C1, 2, 9, 11 220 μF aluminum
  electrolytic

C3, 5, 6, 7 0.1 μF coupling
C4 220 pF ceramic
C8 0.001 μF ceramic bypass
C10 0.1 μF ceramic bypass
C12, 13 10 μF aluminum electrolytic

Resistors
R1 1.2 K
R2, 3, 4 10 K
R5 100 K audio-taper pot w/switch
R6, 12 100
R7, 9, 10 1 K
R8 330 K
R11 10 K audio-taper pot w/switch
R13 10
R14 4.7 K
R15 1 M
R16 470

Semiconductors
D1, 2 1N4007
D3, 4 1N914 or 1N4148
Q1, 3 2N3904 NPN transistor
U1 MAX410 or TL071 or LT1097 op
  amp
U2 4N25 optoisolator
U3 LM386 audio power driver

Miscellaneous
S1 SPST switch (integral to R5)
S2 SPST switch (integral to R11)
T1 10 K:600 ohm transformer (Mouser
  p/n 42TL019 or equivalent)
9 V batteries, printed circuit board,
  solder, wire, etc.

HWR Schematic

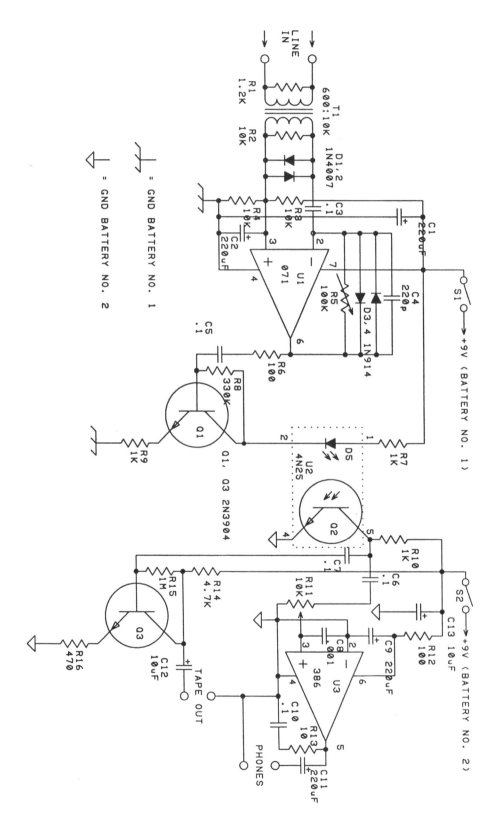

U1 output couples though R6 and C5 to base of Q1, configured by R7, R8, and R9 as a driver for D5, an LED integral to optoisolator U2.

C1 and C2 decouple the first supply.

Phototransistor Q2 is integral to U2, biased by R10. Q2 output is taken off the collector and couples through C6 to volume-control pot R11, whose wiper couples to noninverting input of U3. C8 shunts RF at U3 input; R13 and C10 form the standard snubber. Audio couples through C11 to headphone output.

Q2 output also couples though C7 to base of Q3 configured by R14, R15, and R16 as a common-emitter amplifier with gain of 10. Output is taken off the collector. Signal couples through C12 to tape output.

R11, C9, and C13 decouple the second supply.

DETAILS

Given two independent supplies, both S1 and S2 must be ON for the unit to work. R5 controls preamp gain, R11 controls headphone volume.

The 600-ohm balanced line has another name: telephone line. It conducts line-level audio many miles with trivial loss. In fact, any unshielded, twisted pair of 20- to 30-gauge copper wire will serve as transmission line. This includes the ubiquitous spare pair in the telco trunk. With careful shopping a mile of 24-gauge twisted-pair cable can be had for less than four bills.

The HWS is a lot more versatile than the board-mounted mike implies. The mike could mount in a hose or a corner; become part of a spike setup, etc. The system also has nonsurveillance uses. The HWS's low current drain suits it to hardwired listening posts at tactical points on vast stretches of land. Its tremendous sensitivity invites use as an area-coverage mike, as in a corner boundary.

Inexplicably, hardwire remains a well-kept secret. Properly installed, it defeats 99 percent of debugging sweeps. No other technique transmits audio farther, with greater clarity, on such flimsy current. The budding operator might test this system with a mile of wire to give himself an earful of pro-quality hardware in action. Across the street or across town, hardwire makes a great solution.

THE FLIPSIDE

Hardwire creates a direct link between the target and the operator. To be caught

HWR Circuit Board

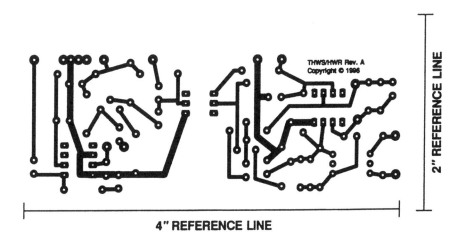

2" REFERENCE LINE

4" REFERENCE LINE

HWR Stuffing and Wiring Diagram

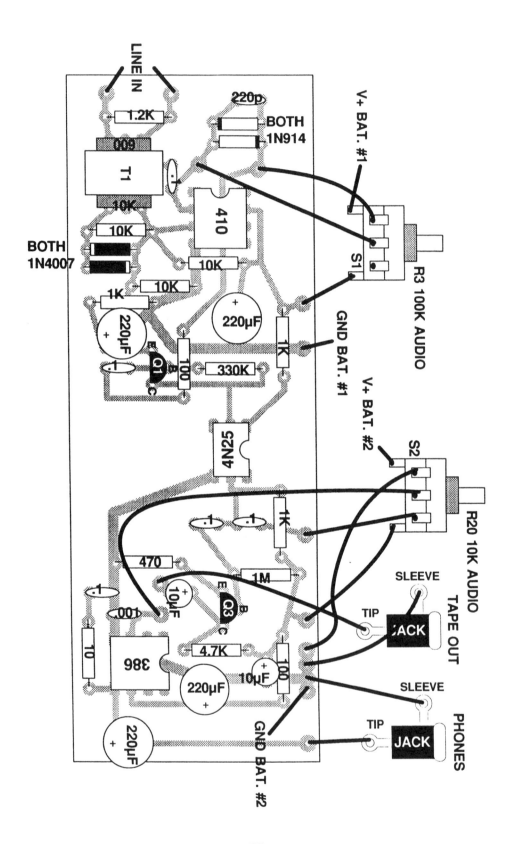

HWR Prototype

manning the post means instant conviction or an indefinite hitch in the Snitch Corps. Should the target discover the wire specialist's Strange and Twisted Pair, he may decide to give it a lethal jolt, as from a phone-line tap-zapper, or simply plug the pair in the nearest AC outlet. In fact, all kinds of ugliness could surge through the line. This explains the HWR's use of an optoisolator. The manufacturer rates its isolation better than 2 kV. To keep this protection the two electrical systems *must* remain physically and electrically isolated. Mount the device in a 100 percent plastic case, use insulated standoffs, mount the two batteries and the two pots apart from each other, use plastic knobs. Don't touch the "hot" end of the piece while it's connected to the line.

These (extreme, to some) safeguards will not thwart all hazards. Accidental contact with a high-tension power line, a lightning bolt, etc. could cause the operator serious harm or death. The pair should run through a UL-certified lightning arrestor, and even so, be disconnected during a thunderstorm.

Remember: Anyone who elects to build and/or use a project assumes *all* risks.

"BRIEFCASE" OPTIBUG

Situation: Hastily called meeting that must be recorded from a location in view of the premises. For whatever reason RF is not an option, nor will the agent be able to optimize transmitter placement in the room.

The gig calls for an optical transmitter, small enough to be built into a briefcase or a handbag. It must contain an extremely sensitive audio stage driving an IR emitter powerful enough to make alignment with the receiver uncritical. In fact, the receiver should be able to work off room scatter. It will have an operational life of no more than a few hours, making battery power practical.

Solution: "Briefcase" optibug (BRO).

CIRCUIT DESCRIPTION

Microphone M1 gets its bias through R1. M1 output couples through C3 to noninverting input of U2-a, biased through R4. U2-a is configured as a noninverting amp with gain determined by ratio of R5 to the net impedance of the network formed by R3, C2, and C4. D1 and D2 clip the output at ~0.7 Vp-p; C5 rolls off response above 7 kHz.

U2-a output passes through lowpass network formed by R6 and C6, coupling through C7 to the modulation port of U2, configured as an astable multivibrator by R7 and R8, whose free-running frequency is set at ~60 kHz by C9. C8 serves as an ultrasonic shunt at pin 5 of U1.

U2-b is configured as a DC voltage

BRO PARTS LIST

Capacitors
C1, 10, 13 220 μF aluminum electrolytic
C2 , 3 10 μF aluminum electrolytic
C4, 14 2.2 μF aluminum electrolytic
C5 220 pF ceramic
C6 0.022 μF 20 percent or better
C7 0.1 μF coupling
C8 0.001 μF ceramic bypass
C9 0.001 μF poly, 5 percent or better
C11 0.01 μF ceramic bypass
C12 1,000 μF aluminum electrolytic

Resistors
R1, 7 2.2 K
R2 200
R3 100
R4 22 K
R5 100 K
R8, 9, 10 10 K
Rx 3-watt metal oxide (see text)

Semiconductors
D1, 2 1N34A germanium diode
D3-34 IR LED (see text)
Q1-4 IRLD024 power MOSFET
U1 LMC555 CMOS timer
U2 LM833 or TL072 or LF353 or
    NE5532 dual op amp

Miscellaneous
J1, 2 jumpers
M1 electret condenser microphone (Radio Shack 270-090 or similar)
Power supply, printed circuit board, solder, wire, etc.

BRO Schematic

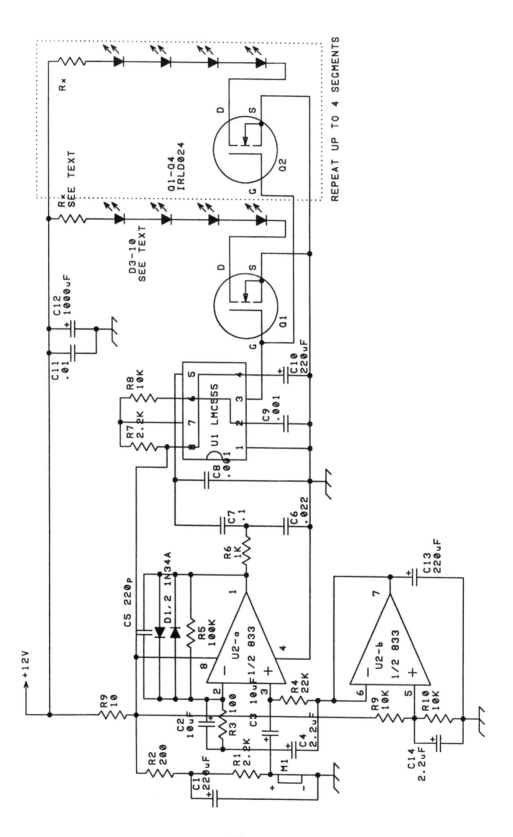

BRO Circuit Board

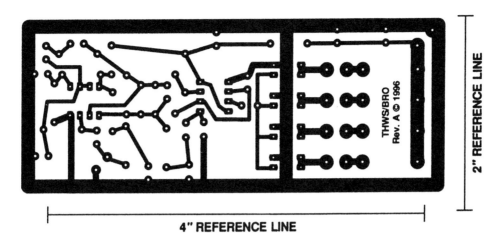

4" REFERENCE LINE

2" REFERENCE LINE

follower whose input is biased at ½V+ by divider R9-R10. U2-b output provides a stable bias reference for U2-a.

U1 output is a frequency-modulated squarewave train tied to the gates of power MOSFET's Q1-Q4. Each MOSFET can drive one to eight infrared LEDs wired in series. Rx is a 3 W power resistor whose value is chosen to limit LED drive current to the desired level.

R2, R9, C1, C10, C11, C12, C13, and C14 decouple the supply.

DETAILS

The builder must determine the value of Rx, the current-limiting power resistor in series with each IR LED segment. The value depends upon: (1) system supply voltage, (2) number of LEDs per segment, (3) voltage drop per LED, and (4) desired current to flow through that segment. Procedure:

Step 1: Choose the supply voltage (12 V for this example).

Step 2: Choose the type and number of LEDs for the segment (for this example use four Panasonic LN175PA wide-dispersion IR LEDs).

Step 3: Choose the desired current level for the segment (100 ma for this example).

Circuit "A"]

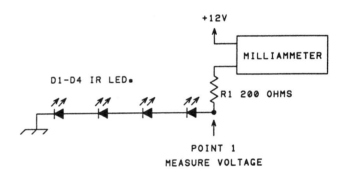

POINT 1
MEASURE VOLTAGE

Step 4: Breadboard circuit "A." Use the actual LEDs that the BRO will drive. Let R1 equal some convenient value that will not pass excessive current, say, 200 ohms. Apply power (12 V), measure the voltage at Point 1. This gives the actual voltage drop across the LEDs, here found to be 5 V. The voltage across R1 equals supply voltage minus voltage drop across diode string. In this case 12 V - 5 V = 7 V.

Step 5: Knowing the voltage across the resistor and the target current of 100 ma, use Ohm's law to calculate the value of Rx to pass that current to the diodes: Rx = 7 V ÷ 0.1 A = 70 ohms.

Step 6: Replace the 200-ohm resistor with a 70-ohm 3 W resistor (68 ohms, a close standard value, is okay). Apply power and verify that current flowing through the segment is acceptably close to the target value.

Step 7: If the measurement revealed a significant difference between predicted and measured current, adjust Rx to deliver the desired current. If too little current flowed, lower the value of Rx; if too much, increase the value of Rx.

Step 8: Run the LEDs at this current for 10 minutes. If they become more than barely warm to the touch the current level is too high; reduce current by 25 percent and re-test.

The procedure shows what DC current flows through the diode string at a given supply voltage and with a given value of Rx. It also reveals the voltage drop per diode (for the LN175PAs, 5 V / 4 diodes = 1.25 V). This value can be used to calculate a new Rx for a lower supply voltage, or for more or fewer diodes of *exactly the same type*. For instance, say the user wants to drive six LEDs off a 10 V supply, at a current of 80 ma. Voltage drop across the diodes is 6 x 1.25 V = 7.5 V. Drop across the resistor = 10 V - 7.5 V = 2.5 V. Resistance that passes 80 ma = 2.5 V / 0.08 A = 31.25 ohms = 30 ohms closest standard value.

Voltage drop varies slightly but significantly among different IR LEDs. The builder can use the procedure given above to derive a value of Rx for any LED/supply voltage combination.

Segments do not have to be identical. The prototype drove four LN175PAs off one segment, four F5D1s off another, each at a different current level.

The builder must observe several limitations. First, a segment cannot contain more LEDs than the supply can drive. The maximum number of LEDs is that which produces a voltage drop at least one volt below the supply voltage. For a 12 V supply, using LEDs that drop 1.3 V apiece, eight should be the limit (drop = 8 x 1.3 V = 10.4 V). An 8 V supply would power only

five of these LEDs.

Second, the power per segment must not exceed maximum supply voltage x maximum current = 12 V x 200 ma = 2.4 watts. Each Rx must be rated 3 watts and be of flame-resistant metal-oxide type, and should stand about ½ inch off the board. The conservative power rating is chosen to avoid excessive (and hazardous) component heating.

Third, the drive level must not exceed the rated current capacity of the LEDs, or they may be destroyed. The procedure above fixes continuous current; the BRO drives the LEDs with a squarewave having an approximate 50 percent duty cycle, meaning the IR LEDs can be pulsed at about twice their maximum rated continuous forward current. A Radio Shack 276-143 "high output" IR LED is rated 20 ma continuous forward current. The BRO should be able to drive a string of them at ~38 ma without ill effect. (The builder would have to verify this empirically, however.) Pushing the LED current limit can defeat itself, for LEDs lose efficiency as their temperature rises.

Supply voltage should not exceed 12 V to avoid damage to the C555, nor should it drop below about 7 V to avoid shutting down the preamp.

The LEDs mount on their own separate board. The 7-LED circuit board is shown as an example. The builder should lay out his own board to suit the application, with an eye toward using copper foil to heatsink the LEDs.

LED Array Circuit Board

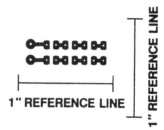

BRO Stuffing and Wiring Diagram

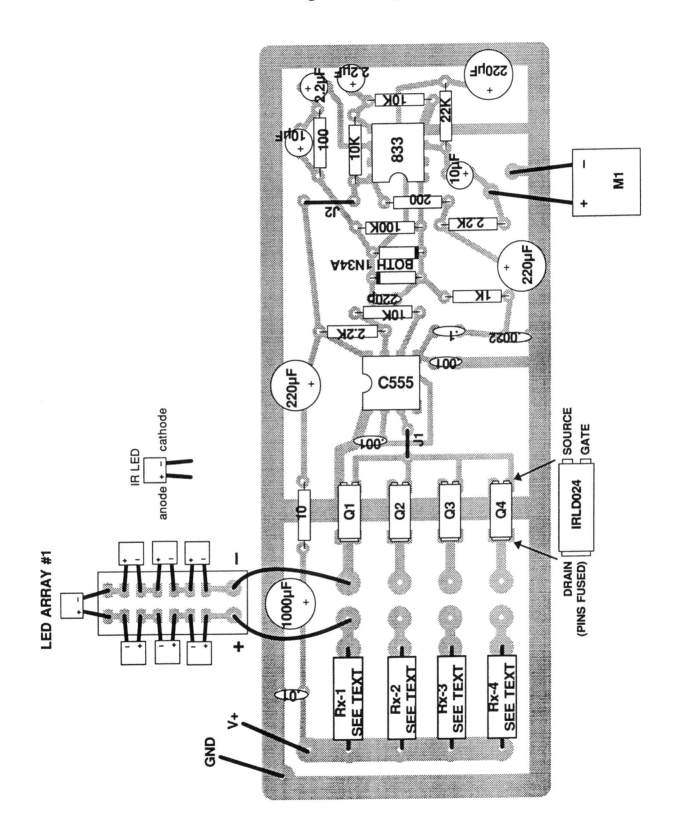

BRO Prototype

The BRO can drive as few as one or as many as 32 IR LEDs. Driving 32 diodes at 200 ma each generates an extremely powerful IR signal. LEDs with integral lenses—such as the F5D1 or the Radio Shack 276-143—emit a signal that some receivers can pick up from across the street, without external optics.

Briefcase/handbag mounting options barely tap the unit's potential. For instance, the BRO could be built into a weatherproof case and mounted under the rear bumper of an automobile, the mike placed inside the car, giving a mobile IR bug that doubles as a tracking beacon: to a camcorder, IR LEDs light up like Times Square.

Although 60 kHz isn't RF, squarewaves are rich in low-RF harmonics. These could trigger a sniffer. To minimize emissions, wiring should be kept short and neat. For board-array separation greater than a few inches, use RG174/U coaxial cable.

The builder should select a power supply that can deliver the required current. Maximum rated power eats 800 ma and change. The user should burn in his final BRO configuration under careful observation for overheating at any site. An accidental short of a high-current battery could easily start a fire or rupture the battery. Appropriate wiring techniques should be applied.

A single, very-high-power LED serves some gigs best, but the builder must choose an IR emitter with care. Some high-power types will take all the current the BRO can provide, but they lack the speed to respond at 60 kHz. They demodulate the FM and act as though driven by an intensity modulator.

"POSTAGE STAMP" OPTIBUG

Low-power FM infrared transmitter, less than 1 inch on a side.

CIRCUIT FUNCTION

Microphone M1 gets its bias through R1. M1 output couples through C1 to base of Q1 configured as a common-emitter amplifier by R2 and R3. C2 rolls off response above 7 kHz. Q1 output couples through C3 to pin 5 of U1, a CMOS 555 timer configured as an astable multivibrator by R4 and R5. C4 sets carrier frequency at ~60 kHz. U1 drives D1 directly. C5 decouples the supply.

DETAILS

Use miniature caps to minimize the "postage stamp" optibug's (PSO's) vertical profile: C1 and C3 should be Mouser p/n 581-UEZ-104K1 (or equivalent); C2 Mouser p/n 581-UEC220J1; C4 Mouser p/n 581-UEC102J1. C5 should be tantalum, 10 WVDC. Resistors should be 1/8 W 5 percent. Press Q1 close to the board before soldering. M1 is a low-profile mike whose pin polarity allows it to sit flush with the board and hang off the edge. The Radio Shack 270-090 has an opposite pin arrangement and will have to extend above the other parts or bend away from the board or solder to the foil side.

The low power level dictates an IR LED with an integral lens, such as an F5D1 or a Radio Shack 276-143 (prototype photo shows wide-dispersion LED in place).

The PSO was designed to run off a 3 V lithium coin cell (e.g., Panasonic CR2477-1HF), soldered to pads on the foil side of the board. The prototype is fitted with a coin-cell holder not practical for field work. The PSO running at 3.0 V drew 2.8 ma. It will run below 2.0 V, with a drop in optical output. Predicted life on a 1,000-mah cell: ~2 weeks. Optical output and current drain rise with supply voltage. A practical limit looks to be 6 V.

PSO PARTS LIST

Capacitors
C1, 3 0.1 μF coupling
C2 22 pF ceramic
C4 0.001 μF
C5 10 μF tantalum

Resistors
R1 4.7 K
R2 1 M
R3, 5 10 K
R4 2.2 K

Semiconductors
D1 infrared LED (see text)
Q1 2N3904 NPN transistor
U1 TLC555 CMOS timer

Miscellaneous
M1 Panasonic WM52BM electret condenser microphone
Printed circuit board, lithium coin cell, etc.

PSO Schematic

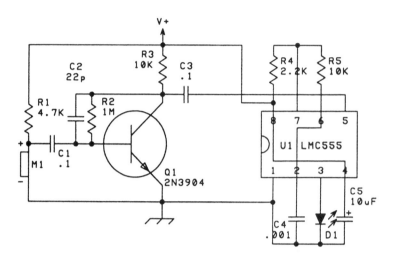

Stuffing and Wiring Diagram

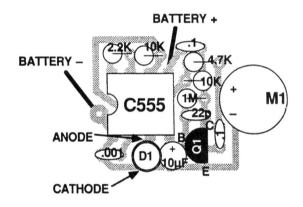

Foil Side of Board

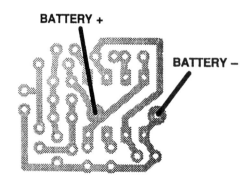

PSO Prototype

PSO Circuit Board

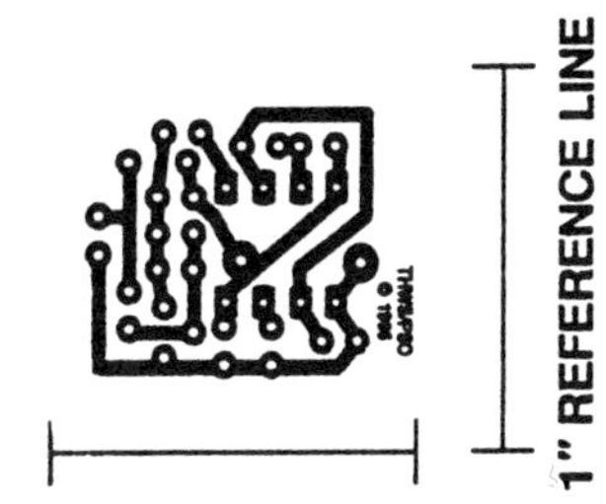

The completed device with power supply is about the size of a postage stamp and maybe half again as thick. It is meant to mount in a window with LED and mike facing the pane. A corner is the preferred locus, as this takes advantage of the boundary effect to boost the audio by 18 dB. A typical quickie mounting method entails lumping adhesive (caulk, "liquid nails," etc.) atop U1 and pressing the works into the corner of a window.

VARIATION ON A THEME

The PSO will also work with a bipolar 555. Keeping the 3.0 V supply, current draw rises to 22 to 45 ma, depending upon the LED driven. This dictates a stronger power supply, as few lithium coin cells are rated more than 15 ma continuous drain. Bypass cap C5 should be increased to 100 uF. The carrier drops to ~45 kHz but can be restored to 60 kHz by lowering the value of C5. The carrier eases back up toward 60 kHz as the supply approaches 5.0 V. With no series resistor for D1, LED current is a function of supply voltage, which must be tempered by thermal limits of D1 and U1. The bipolar version gives several times the range of the CMOS version.

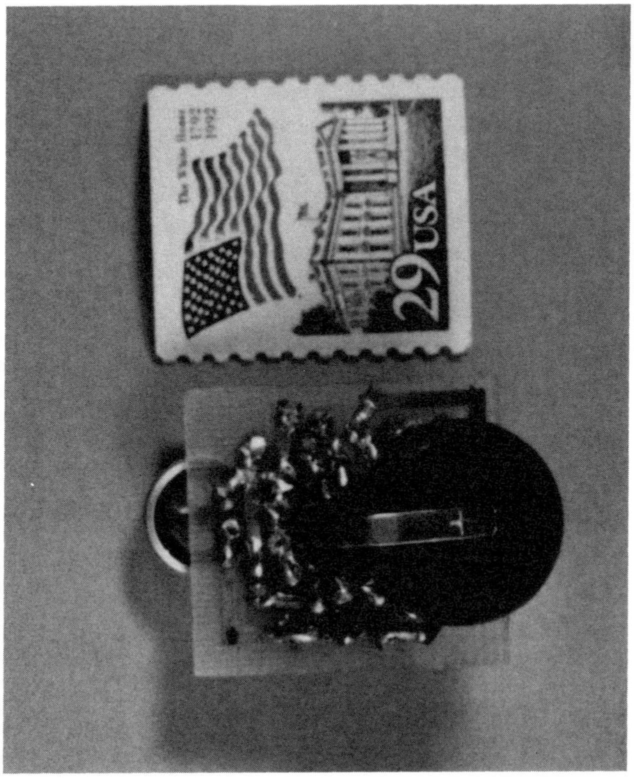

PSO Foil Side

OPTIBUG RECEIVER

FM optical receiver for use with BRO and PSO.

SYSTEM THEORY

Common phototransistors will respond to 60 kHz, but their relatively slow tR causes them to act as slope detectors for a frequency-modulated 60-kHz carrier. The optibug receiver (OBR) recovers audio by rectifying and lowpass filtering the carrier.

CIRCUIT FUNCTION

Infrared phototransistor Q1 gets its bias selectably through R4 or tank L1-C2 by way of SPDT switch S1. Q1 output couples through C3 to input of U1-a configured by R5, R7, and C4 as a noninverting amplifier with gain variable by pot R6 from 0 to 60 dB.

U1-a output couples through C5 and R8

OBR PARTS LIST

Capacitors
C1, 6, 14, 18, 20 220 μF aluminum electrolytic
C2 0.0033 μF 5 percent poly
C3, 4 2.2 μF aluminum electrolytic
C5, 16 0.1 μF coupling
C7, 11, 12 0.01 μF, 10 percent or better
C8, 9 0.001 μF, 10 percent or better
C10 0.1 μF, 10 percent or better
C13 0.01 μF ceramic bypass
C15 10 μF aluminum electrolytic
C17 0.001 μF ceramic bypass
C19 0.1 μF ceramic bypass

Resistors
R1, 10, 13 2.2 K
R2 100 K linear-taper pot
R3, 18 200
R4 33 K
R5, 21 10
R6 10 K linear-taper pot
R7 330 K
R8, 9, 16, 17 10 K
R11, 12, 14, 15 22 K
R19 10 K audio-taper pot w/switch
R20 100

Semiconductors
D1 IR LED (see text)
D2, 3 1N914 or 1N4148
Q1 IR phototransistor (see text)
U1 MAX412 (preferred) or TL072 or LM833 dual op amp
U2 MAX414 or TL074 or LF444 quad op amp
U3 LM386 audio power driver

Miscellaneous
J1 jumper
L1 2.7 mH variable inductor, Digi-Key p/n TK1708
S1 SPDT switch
S2 SPST switch (part of R19)
9 V battery, printed circuit board, solder, wire, etc.

OBR Schematic

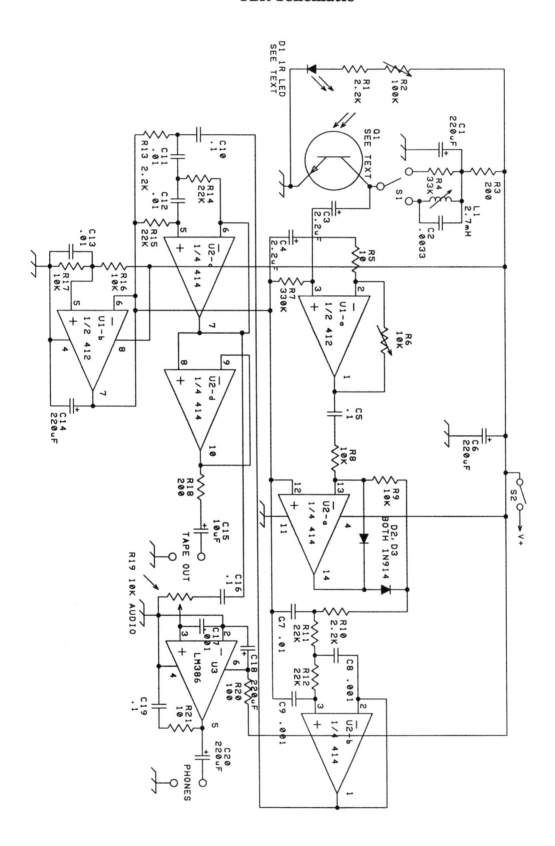

OBR Circuit Board

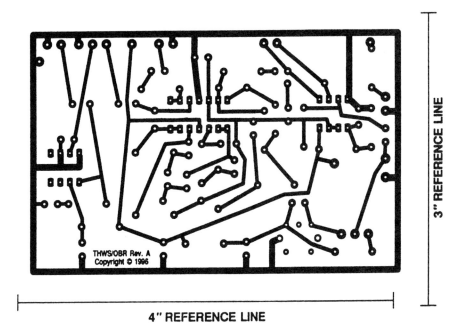

THWS/OBR Rev. A
Copyright © 1996

3" REFERENCE LINE

4" REFERENCE LINE

to U2-a, configured by D2, D3, and R9 as a precision halfwave rectifier.

U2-a output couples directly to U2-b, configured by R10-12 and C7-9 as a quasi-18 dB/octave lowpass filter, cutoff above 7 kHz. U2-b output couples directly to U2-c, configured by R13-15 and C10-12 as a quasi-18 dB/octave highpass filter, cutoff below 700 Hz.

U2-c output couples directly to U2-d, configured as a voltage follower. Signal couples through R18 and C15 to tape output.

U2-c output also couples through C16 to volume control pot R19, whose wiper couples to the noninverting input of U3. C17 shunts RF at U3 input; R21 and C19 form the standard snubber. Audio couples through C20 to headphone output.

U1-b is configured as a voltage follower whose input is biased at ½V+ by divider R16-17. U1-b output serves as a stable bias reference for U1-a and U2-a/-c.

Pot R2 in series with R1 forms a variable intensity control on D1. Its function is explained below.

C1, C6, C13, 14, 18, and R20 decouple the supply.

DETAILS

The prototype used a Radio Shack 276-145 infrared phototransistor. Other types with similar speed (tR/tF 5 to 10 μsec) will also work, but require a transparent plastic case to be ditherable by D1.

To align tank L1-C1, supply a 60-kHz FM IR source (from, e.g., the BRO, PSO, or ROS). Tune L1 for peak carrier at the output of U1-a, then rotate L1 tuning screw 180 degrees clockwise. This step can also be performed by ear, adjusting L1 for best audio while receiving an audio-modulated IR signal.

Maximum system gain in excess of 85 dB means that the user can run either the preamp or the audio amp wide open, but not both at the same time, as this usually leads to oscillation. The excess of gain was built into the circuit to accommodate various types of intercepts. Some favor full audio gain combined with medium preamp gain; others, the reverse; still others peak with preamp and audio amp gain set about halfway. The user is encouraged to experiment with various gain combinations.

The resistor-biased Q1 generates a

OBR Stuffing and Wiring Diagram

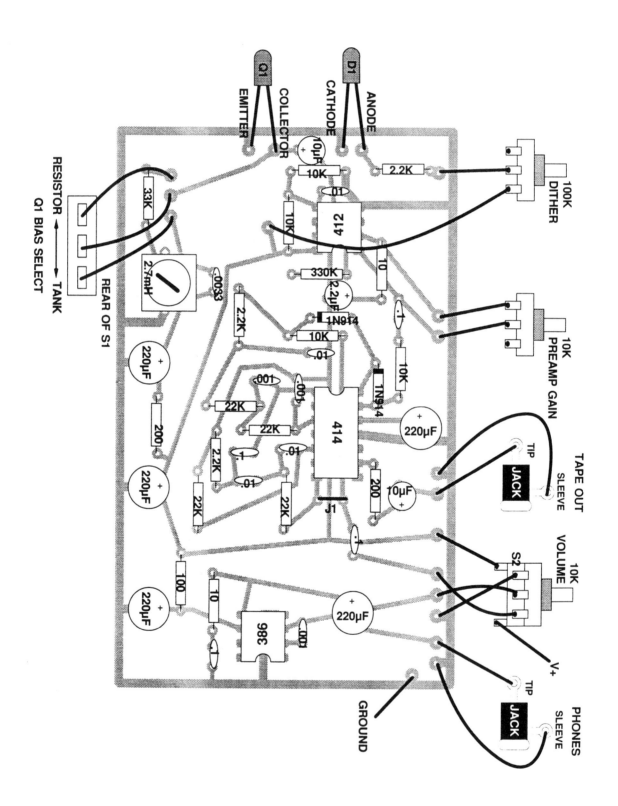

OBRPrototype

healthy 60-Hz buzz when looking at most any artificial light source. Hum is reduced greatly when Q1 is biased through a tank resonant near the carrier, with some sacrifice in sensitivity. Beware that switching S1 injects a loud pop into the audio.

Phototransistors depend somewhat on ambient light to bias them in a partly ON state that gives good sensitivity. Sensitivity in low-light settings can be enhanced by local IR source D1, whose intensity is variable by R2. D1 should rest directly behind Q1, held in place by a dab of glue or a strip of tape. Optical dithering has been described elsewhere, e.g., *Radio Electronics* (May 1989, p. 41-43). Dithering should be used only when needed, because it consumes current and adds noise.

BEYOND DETAILS

The operative word is *optical*. Good optics can make a mediocre system sound superb; poor optics can neutralize a great system.

This phase of optibugging knows a dearth of detail because optical trade secrets cost more to acquire than book royalties can pay. The serious student just wading into the mode might undertake systematic experimentation, working up in optical complexity and distance; testing at night and during the day, with and without strong 60 Hz interference. Tips:

- Start with a single lens, 2 to 3 inches in diameter, biconvex or planoconvex. (If those terms mean nothing to you, read a book on optics.)
- Next, move up to simple compound optics, say, a cheap rifle scope.
- Realize that lenses work at the transmitter, too, and that a collimated beam can double system range.
- Think Fresnel lenses.
- Remember that a phototransistor's optical axis does not always align perfectly with its physical axis.
- An aluminum parabolic dish is just about ideal for focusing infrared energy.
- Phototransistors can run in parallel.
- Other things equal, photosensors with integral lenses give greater range than those without integral lenses.
- The wire specialist will find some form of reference optical source (ROS), an invaluable aid to serious research (see Chapter 9).

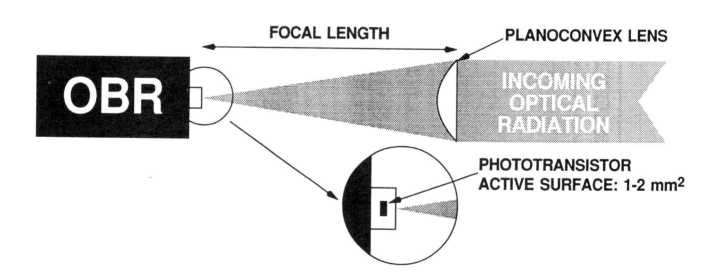

OBR: Close-Up of D1 Behind Q1

REFERENCE OPTICAL SOURCE

The true test of an optical receiver lies in its performance at long range, under real-world conditions. Yet proper field-testing is difficult, often impractical.

An alternative to open testing is the reference optical source, a portable generator of frequency-modulated infrared radiation, whose optical output, frequency deviation, and carrier frequency are selectable and, just as important, reproducible.

CIRCUIT DESCRIPTION

Op amps U1-a and U1-b are configured by R3-R9, D1, D2, C3, and C4 as a controlled-amplitude sinewave oscillator adapted from the Wien oscillator described in *IC Op-Amp Cookbook*, 3rd Ed. (p. 450-1). Frequency is ~1,300 Hz, amplitude is constant at ~2.9 Vp-p. Sinewave output couples through R10 to S2, a SP4T switch that selects a companion resistor to divide the signal.

Output of the attenuator network couples through C5 to pin 5 of U2, a CMOS 555 timer configured as an astable multivibrator by R15 and R16. S4 selects between two timing capacitors; C9 gives ~60 kHz; C8 is trimmable to 455 kHz and can be padded with C10 if needed to cover 300 to 700 kHz. U2 output drives IR LED D3 through current-limiting resistors R17-20 one at a time, selected through SP4T switch S3. C7 shunts the carrier at pin 5.

Three-terminal 5 V regulator U3 supplies U2.

ROS PARTS LIST

Capacitors
C1, 2, 6 220 μF aluminum electrolytic
C3, 4 0.1 μF poly, 5 percent or better
C5 10 μF aluminum electrolytic
C7 0.01 μF ceramic bypass
C8 10-50pF trimmer
C9 0.001 μF poly, 5 percent or better
C10 22 pF ceramic (optional—see text)

Resistors
R1, 2, 11, 16, 20 10 K
R3, 4, 6, 9, 15 2.2 K
R5 750
R7 51
R8 51 K
R10 30 K
R12, 18 1 K
R13 100
R14 10
R17 200
R19 4.7 K

Semiconductors
D1, 2 1N914 or 1N4148
D3 Radio Shack 276-143 IR LED (circuit can use any common IR LED)
U1 TL072 or TL082 or LF353 dual op amp
U2 LMC555 timer
U3 LM78L05 5 V regulator

Miscellaneous
S1 SPST switch
S2, 3 SP4T switch
S4 SPDT switch
9 V battery, wire, circuit board, etc.

ROS Schematic

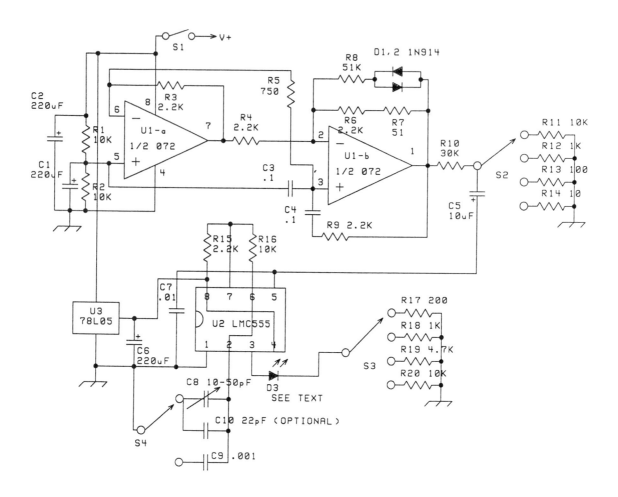

ROS Circuit Board

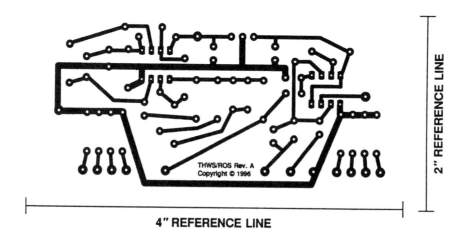

2" REFERENCE LINE

4" REFERENCE LINE

THWS/ROS Rev. A
Copyright © 1996

50

Voltage divider R1-R2 gives a ½V+ bias for U1-a.

C1, C2, and C6 decouple the supply.

DETAILS

Power supply can run as high as 15 V, but should not drop below 7.0 V to avoid shifts in sinewave amplitude.

Power up the ROS, select S4 "high frequency," trim carrier to 455 kHz (or whatever frequency the user desires). If C8 will not tune U2 to 455 kHz, pad it with C10, a 22 pF temperature-stable cap. Using a handy FM optical receiver such as the OBR, confirm function of the intensity attenuator and deviation attenuator.

Attenuator at highest position puts approximately 730 mV$_{p-p}$ into the C555. This strong level is not typical of real-world intercepts but can aid in aligning insensitive receivers. Second position feeds in ~90 mv, more typical of a strong intercept. Third position injects ~10 mv, typical of an average intercept. The fourth position lets only ~1 mv into U2, typical of a relatively faint intercept. The builder can choose new values for R11-14 during construction, using Ohm's law to calculate the amplitude of the resultant feed into U2.

The ROS will work with a bipolar 555 but will not make it much above 220 kHz in this configuration. Also, the user must take special care in selection of current-limiting resistors R17-20 because the bipolar chip will switch up to 200 ma, enough to kill an LED or roast a resistor, and send U3 into thermal shutdown.

The ROS facilitates several tasks:

- It lets the engineer determine conversion gain of various types of FM detectors. Audio feeding the C555 is known, constant, and reproducible. It allows objective benchmarks to be derived for different detectors under standard conditions.
- It helps tune optical receivers for peak audio. The low-distortion sinewave lets the user optimize the tuning of, say, a quadrature detector from the shape and amplitude of a scope display.
- It lets the user test different optical sensors with identical signals.
- It gives a handy way to test a receiver's ability to reject optical interference. This takes the form of powerful 60 Hz energy thrown off by incandescent lights and sodium vapor streetlights. Place the ROS in the glare of, or behind, an incandescent lamp.
- It allows comparison of receivers working at 60 kHz against those working at 455 kHz.
- It can test different LEDs under identical drive conditions; replace D3 with a socket for quick change.
- It can aid alignment of sensor and optics.
- By measuring LED beam angle close in, the operator can use trig to project it over distance. Applying the inverse-square law and atmospheric absorption factors, the operator can derive a useful estimate of range.

And more. Every professional has something like the ROS in his lab.

ROS Stuffing and Wiring Diagram

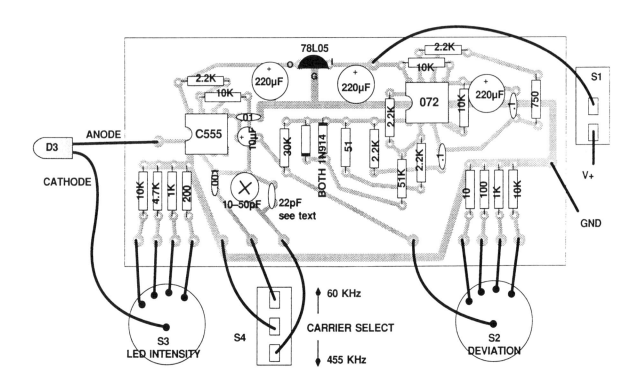

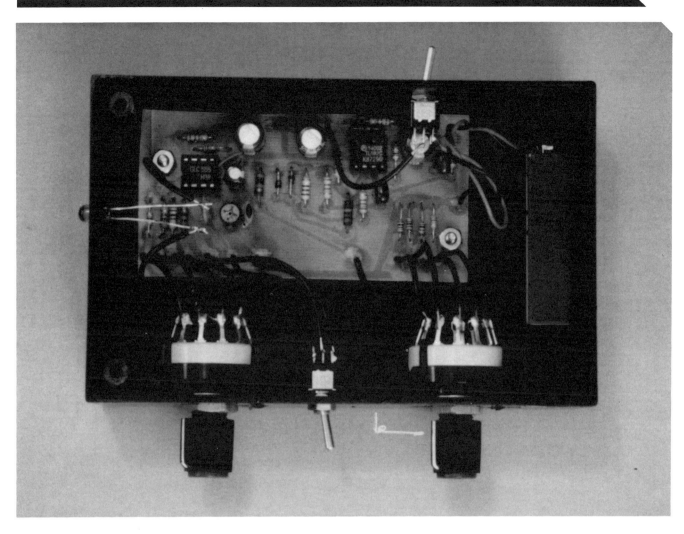

ROS Prototype

FIBEROPTIC REPEATERBUG

Modern bugging signals find themselves having to travel in decreasingly detectable form. RF and IR, and lately ultrasonics and copper wire, debug with relative ease if the sweeper takes the trouble to look for them. Properly hidden fiberoptic cable can be extremely difficult to find. The fiberoptic repeaterbug (FRB) conducts audio first through fiberoptic cable, then shifts to a balanced transmission line.

Due to losses in the particular cable used here, fiberoptic transmission is practical only over some tens of meters. At that point the signal enters an extremely simple *repeater*. A repeater is a combination receiver/retransmitter that might or might not alter the transmission medium. This

FRB TRANSMITTER PARTS LIST

Capacitors
C1, 2 10 μF aluminum electrolytic
C3 1 μF aluminum electrolytic or
   tantalum
C5, 6 0.1 μF coupling
C4 22 pF ceramic

Resistors
R1 100
R2, 4, 5 10 K
R3 1 M
R6 100 K
R7 1 K

Semiconductors
D1 Motorola MFOE71 IR LED built into

fiberoptic coupler
Q1, 2 2N3904 NPN transistor

Miscellaneous
M1 electret condenser microphone
(Radio Shack 270-090 or equivalent)
Fiberoptic cable, 1,000 u core plastic
   fiber (DuPont OE1040 or Eska
SH4001)
Battery, wire, circuit board, etc.

FRB REPEATER PARTS LIST

Capacitors
C1, 3 0.1 μF coupling
C2 22 pF ceramic
C4 10 μF aluminum electrolytic or
   tantalum

Resistors
R1 3.9 K
R2 1 M
R3, 4 10 K
R5 1.2 K

Semiconductors
Q1 Motorola MFOD72 IR
   phototransistor built into fiberoptic
   coupler
Q2 2N3904 NPN transistor

Miscellaneous
T1 600:10 K transformer (Mouser p/n
   42TL019 or equivalent)
Battery, wire, circuit board, etc.

repeater drives a balanced transmission line similar to the one used by the hardwire senderbug. The final receiver is the hardwire receiver. This chain can maintain superb audio over a range of some miles.

TRANSMITTER CIRCUIT FUNCTION

Microphone M1 gets its bias through R2. M1 output couples through C3 to base of Q1, configured as a common-emitter amplifier by R3, R4, and R5. C4 tames the treble emphasis produced by C6. Q1 output is taken off the collector and couples through C5 to the base of Q2, configured as a common-emitter amplifier by R6 and R7, with fiberoptic LED D1 in series with R7. In addition to driving D1, Q1 supplies audio gain.

R1, C1, and C2 decouple the supply.

REPEATER CIRCUIT FUNCTION

Fiberoptic phototransistor Q1 gets its DC bias through R1. Q1 output couples through C1 to base of Q2, configured by R2 and R3 as a common-emitter amplifier whose collector load consists of the 10 K winding of T1 loaded by R4. C2 takes the edge off the treble emphasis produced by C3. The 600-ohm winding of T1 is loaded by R5. It couples directly to a balanced transmission line.

C4 decouples the supply.

FRB Transmitter Circuit Board   **FRB Transmitter Schematic**

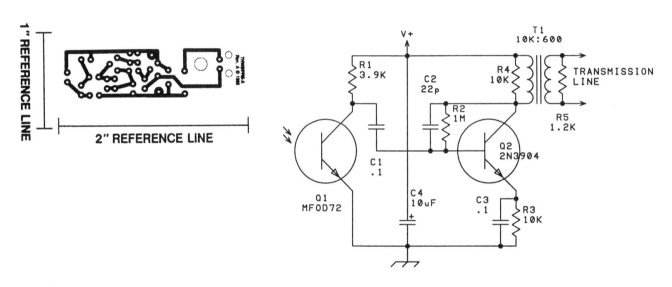

FRB Transmitter Stuffing and Wiring Diagram

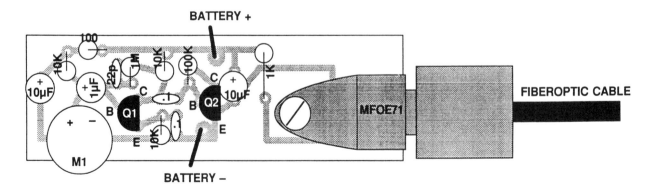

56

FRB Transmitter Prototype

FRB Repeater Schematic

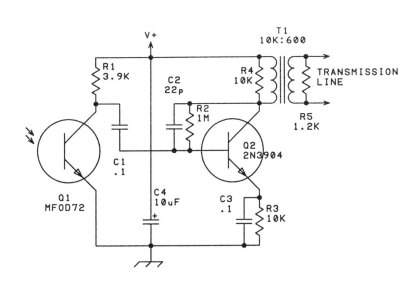

FRB Repeater Circuit Board

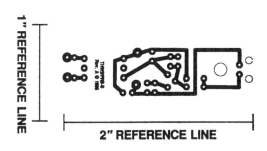

57

FRB Repeater Stuffing and Wiring Diagram

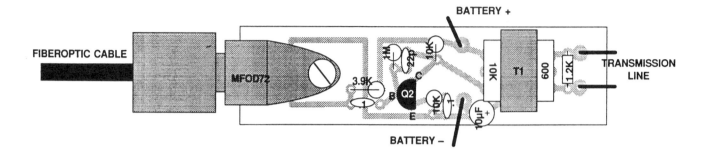

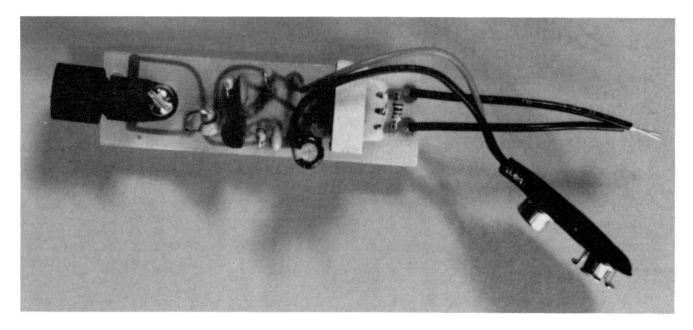

FRB Repeater Prototype

DETAILS

The FRB transmitter will run off ~2.5 V to 15 V. Prototype drew 1 ma at 3 V, 2.5 ma at 5 V, 4.3 ma at 7.5 V, 6 ma at 10 V, 9.7 ma at 15 V. Sensitivity and optical output rise with supply voltage. It is apparent from inspection of the LED driver circuit that the builder can alter the value of R7 (which may necessitate changing R6) to increase or decrease optical output at a given supply voltage. The nominal values of R6 and R7 give a good compromise between power drain and range. According to Motorola data, incremental LED output per milliamp of drive drops sharply above 10 ma.

The FRB uses 1,000 jacketed plastic fiber. Transmitter-repeater range varies with transmitter supply voltage and value of R7, but tops out around 200 feet.

The FRB repeater is designed to run on a 9 V battery. It will function over the range 3 to 12 V, but shifting the supply much below 6 V or above 10 V will demand proportionate increase or decrease of the value of R1.

PROJECT NO. 11

PIEZO PROBE

Piezoelectric transducers are extremely efficient, horns are directional, and common piezo horn tweeters made to respond above 5 kHz show a stout midrange cavity resonance when used as microphones. Their output equals that of electret condenser mikes without electrets' self-noise. Their speech-band impedance is relatively low, meaning low thermal noise and low $(I_n{}^XR_S)$ noise. These traits combine with a low-noise/high-gain op amp to make the piezo probe (PZP), a medium-range directional mike system that, literally, can fit in the palm.

CIRCUIT FUNCTION

Piezo microphone M1 couples directly to noninverting input of U1, biased through R2 at ½V+. Op amp gain is determined by the ratio of R3 to the net impedance of the network comprised of R1, C1, and C2. This produces a sharp treble emphasis. C3 rolls off response above 5 kHz. D1 and D2 clip U1 output at about 1.2 Vp-p.

Preamp output couples through C5 to noninverting input of U2-b, configured as an AC voltage follower by R9. Output feeds through R8 and C11 to tape output.

Signal also couples through C4 to R4 which, with volume control pot R5, forms a divider that neutralizes some of U3's 26 dB of gain. U3 output couples through C10 to the headphone jack. R7 and C9 form the standard snubber.

U2-a is configured as a DC voltage follower whose noninverting input is biased

PZP PARTS LIST

Capacitors
C1, 2, 11 10 μF aluminum electrolytic
C3 330 pF ceramic
C4, 5, 6 0.1 μF coupling
C7 0.001 μF ceramic bypass
C8 470 μF aluminum electrolytic
C9 0.1 μF ceramic bypass
C10, 12, 14 220 μF aluminum
    electrolytic
C13 0.01 μF ceramic bypass

Resistors
R1, R6, R8 100
R2 22 K
R3, 9 100 K
R4 68 K
R5 10 K audio-taper pot w/switch
R7 10
R10, 11 10 K

Semiconductors
D1, 2 1N914 or 1N4148
U1 MAX437 op amp
U2 TL072 or LF353 dual op amp
U3 LM386 audio power driver

Miscellaneous
J1 jumper
M1 piezoelectric horn tweeter (see text)
S1 SPST switch (part of R5)
9 V battery, wire, circuit board, mike
    cable, etc.

MM: PZP Schematic

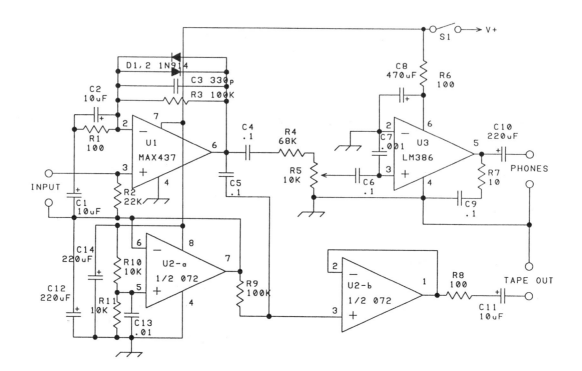

PZP Circuit Board

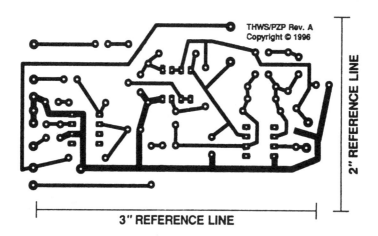

PZP Stuffing and Wiring Diagram

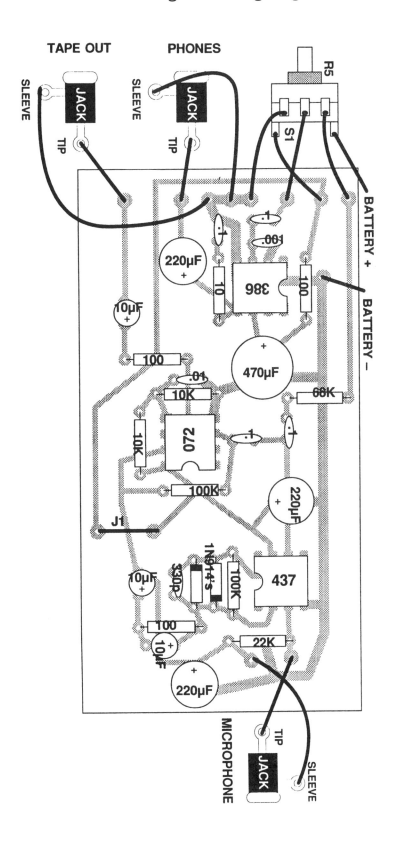

PZP Amp Prototype

at ½V+ by voltage divider R10-R11. U2-a's output provides U1 and U2-b with a stable bias reference.

C8, C12, C13, C14, and R6 decouple the supply.

DETAILS

Plastic piezo horn tweeters similar to the one shown can be had for as little as $4 (e.g., Motorola KSN1005A from Hosfelt Electronics) or as much as $14 (Radio Shack p/n 40-1396).

The mike input is referred to ½ V+, not battery ground. A metal case will short the internal reference unless insulated jacks are used. The prototype used a plastic case.

Due to extremely high gain and relatively high input impedance, the mike jack should mount as close to the board as possible. Wiring should be neat, leads short. High-quality, shielded mike cable should be used and preferably kept shorter than 3 feet.

Most piezo tweeters are made of plastic, leaving them susceptible to AC power fields. Hum did not pose a problem out of doors and away from power lines. For use near

AC fields the wire specialist might house the entire system in a metal box.

Besides the mike, the key to the PZP's fantastic performance is the open-loop gain of the 437: 80 dB as high as 6 kHz, less than 90 dB at 600 Hz (here tempered somewhat by the low supply voltage). The gain/frequency-determining network on the 437 pretty well tops out gain. Other, equally quiet op amps run a good 20 dB cooler.

Initial tests will highlight the futility of using the PZP in all but dead-quiet surroundings. The sound of breathing is enough to clip the audio. Horn output has a tinny tenor that turns out to be a plus, for this resonance lands in a key part of the speech band.

The PZP has at least three uses. As is, it makes a short- to medium-range directional mike whose directionality could be enhanced by a frontal skirt, say, a 10-inch segment of 5-inch PVC pipe. Mated to a parabolic dish 12 to 24 inches it becomes a medium-to-long-range directional mike. Finally, it has a way of cropping up when talk turns to through-the-wall intercepts: a bead of caulk around the horn rim, the mouth sealed against the wall—these are the little gimmicks that warm a wire specialist's icy heart.

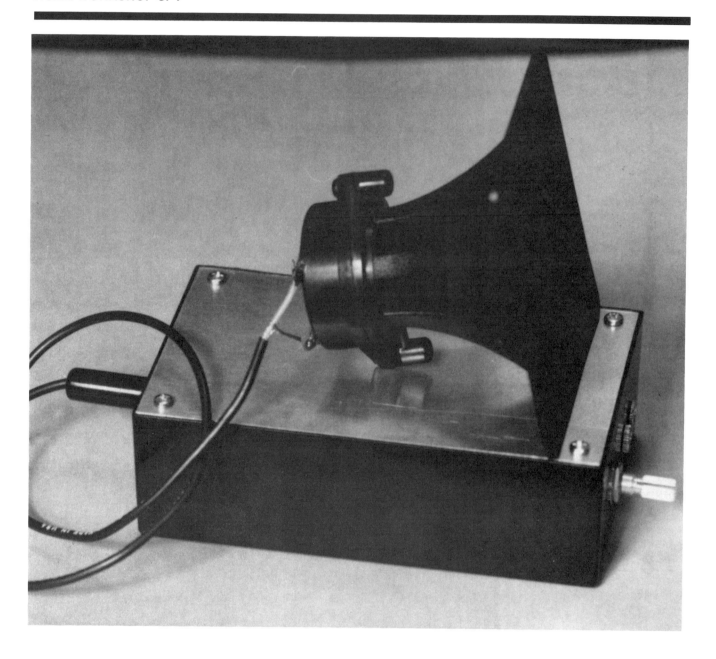

REFERENCE SOUND SOURCE

The reference sound source (RSS) does for directional mikes and room bugs what the ROS does for optical receivers. It generates a clean 1,300 Hz tone at four selectable loudness levels and delivers it constantly or intermittently.

CIRCUIT FUNCTION

Op amps U1-a and U1-b are configured by R3-9, C3-4, D1, and D2 as the same controlled-amplitude sinewave generator found in the ROS. Frequency is ~1,300 Hz. Divider R1-R2 furnishes a ½V+ bias reference for U1-a.

Sinewave audio couples through C5, R10, and R11 to input of U2. S2 selects one of four companion resistors to attenuate the signal level going into U2. C12 shunts RF at U2 input; R17 and C11 form the standard snubber. Audio couples through C7 to the speaker.

U3 is a CMOS 555 timer configured as an astable multivibrator with a duty cycle of ~50 percent and a rate of ~1 Hz. When pin 3 is high, it activates control channel "A" of U4, a 4066 CMOS quad bilateral switch. This turns on the path between pins 1 and 2. Since pin 2 is grounded, the AC signal from the high-impedance juncture of R10-R11 shunts to ground, muting the tone. When U3 pin 3 goes low, the path between U4 pins 1-2 effectively becomes an open circuit, allowing audio to pass through R10-R11 unimpeded. S3 selects constant or intermittent tone. C8 provides an AC ground for U2's FM port.

C1, C2, and C6 decouple the supply.

RSS PARTS LIST

Capacitors
C1, 2, 6, 7 220 µF aluminum electrolytic
C3, 4 0.1 µF poly, 5 percent or better
C5, 11 2.2 µF aluminum electrolytic
C8, 13 0.1 µF ceramic bypass
C9 100 µF aluminum electrolytic
C12 0.001 µF ceramic bypass

Resistors
R1, 2, 18 10 K
R3, 4, 6, 9, 17 2.2 K
R5 750
R7 51
R8 51 K
R10, 11 22 K
R12, 14 100
R13, 17 10
R15 510
R16 1 K

Semiconductors
D1, D2 1N914 or 1N4148
U1 TL072 or TL082 or LF353 dual op amp
U2 LM386 audio power driver
U3 LMC555 CMOS timer
U4 4066 quad bilateral switch

Miscellaneous
S1 SPST switch
S2 SP4T switch
S3 SPDT switch
4-16 ohm speaker
9 V battery, printed circuit board, solder, wire, etc.

RSS Schematic

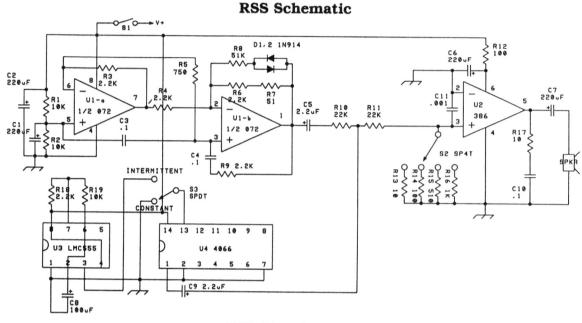

RSS Circuit Board

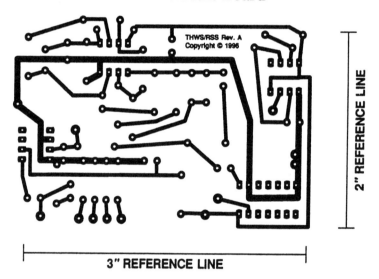

3" REFERENCE LINE

DETAILS

The builder can change the values of R13-16 to customize the four discrete loudness levels. Frequency is alterable by changing C3 and C4 (the two must match to keep sinewave distortion low). Another tuning node is available at R5.

The prototype incorporates a speaker-output jack to accommodate different speakers. Construction particulars, such as case, speaker, etc., mean less than the fact of repeatability from session to session.

Unlike an optical source, a sonic reference must contend with the comb-filter effect, a polytonic cancellation that renders single measurements unreliable, especially indoors. Slight placement changes can shift a reading from reinforcement to cancellation. Averaging can overcome this to some degree. Outdoor testing is best.

The RSS lends a measure of objectivity

to testing mikes, amps, filters, and physical gain tools. It is meant less for measurement than for allowing the tester to correlate a given RSS tone sensitivity with voice sensitivity in the field. By extrapolation, long-range mikes can be tested at

comparatively short distances. One system can be compared with others, even at some later date.

Like the ROS, the RSS should be one of the professional's earliest workshop additions.

RSS Stuffing and Wiring Diagram

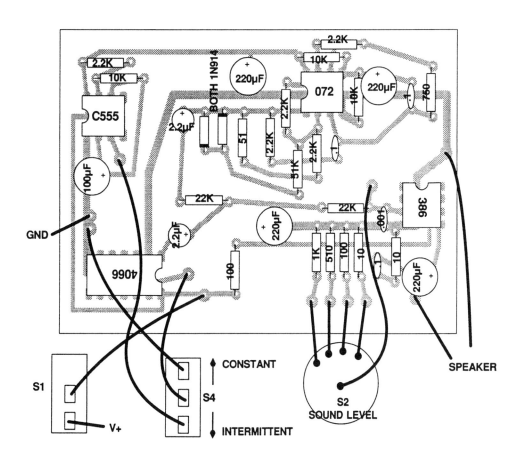

RSS Prototype

PROJECT NO. 13

POCKET PARAMETRIC

Having dealt a hand to the twin gods of high gain and low noise, the pocket parametric (PPM) took the pot by upping their ante with extremely versatile EQ. The PPM will oblige electret condenser, ceramic, and dynamic mikes, and manages its magic off a lone 9-volt.

CIRCUIT FUNCTION

Microphone input couples through C2 to input of U1-a configured as a noninverting amp by R3, R4, R5, and C3. C4 limits high-frequency response; D1 and D2 clip output at ~1.2 V_{p-p}. Note that C2 must be nonpolar, as the DC bias on the microphone side of C2 reverses (compared to ½V+ at U1 pin 3) when changing from a biased electret condenser mike to an unbiased dynamic mike.

PPM PARTS LIST

Capacitors
C1, 6, 8, 9 13 220 μF aluminum electrolytic
C2 10 μF nonpolar electrolytic
C3 2.2 μF aluminum electrolytic
C4 560 pF ceramic, 10 percent
C5, 10 0.1 μF coupling
C7 0.001 μF ceramic bypass
C11 10 μF aluminum electrolytic
C12 0.1 μF ceramic bypass
(Parametric EQ parts not shown on schematic: 2 ea. 0.0033 μF 1 percent poly)

Resistors
R1 200
R2 2.2 K
R3, 11, 15 100
R4 39 K
R5 22 K
R6 10 K audio-taper pot
R7 10 K linear-taper pot
R8 100 K linear-taper dual-ganged pot
R9 100 K
R10 10 K audio-taper pot w/switch
R12, 13 10 K
R14 220 K
R16 10
(Parametric EQ parts not shown on schematic: 6 ea. 10 K 1 percent; 2 ea. 4.75 K 1 percent; 1 ea. 2.49 K 1 percent; 2 ea. 22 K; 1 ea. 470)

Semiconductors
D1, D2 1N914 or 1N4148
U1 MAX412 (preferred) or LM833 or TL072 dual op amp
U3 LM386 audio power driver
U4 MAX412 or LF442 or LF353 or TL072 dual op amp
(Parametric EQ part not shown on schematic: U2 MAX414 or LF444 or LF347 or TL074 quad op amp)

Miscellaneous
S1 SPST switch (part of R10)
S2 SPST switch
Microphone, 9 V battery, printed circuit board, solder, wire, etc.

Preamp output feeds to a parametric equalizer made up of U1-b and quad op amp U2. The basic circuit appears in *Audio IC Op-Amp Applications*, 3rd Ed. (p. 189-191). That circuit has been modified here to allow greater boost/cut and higher maximum Q.

Equalizer output couples through C5 to the divider formed by R9 and volume-control pot R10, whose wiper couples to the noninverting input of U3. C7 shunts RF at U3 input; R16 and C12 form the standard snubber. Audio couples through C13 to headphone output jack.

EQ output also couples off C5 to C10 to U4-a, configured by R14 as an AC voltage

PPM Schematic

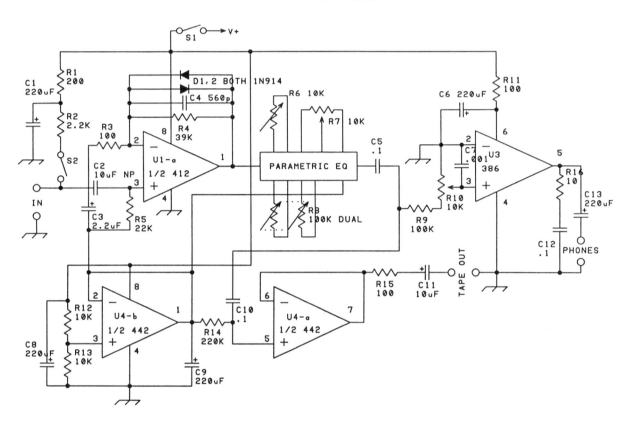

PPM Circuit Board

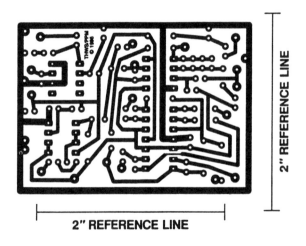

2" REFERENCE LINE

follower. Line-level audio couples through R15 and C11 to tape output.

U4-b is configured as a DC voltage follower whose input is biased at ½ V+ by divider R12-R13. Its output serves as a stable bias reference for the mike preamp, parametric EQ, and tape-output buffer.

An electret condenser microphone, if used, can draw DC bias through R2, switchable in/out by S2.

R1, R11, C1, C6, C8, and C9 decouple the supply.

DETAILS

A pack of filter kings will swallow the PPM circuit board, and, in fact, costly midget pots let the device squeeze into a box that size. The prototype used full-size pots and a box comfortable in a coat-

pocket; the board was given a skirt to facilitate mounting. Microphone-input jack and S2 should mount as close as possible to the board. All leads should be kept short and neat.

Before powering up, set mike bias switch S2 in/out as dictated by choice of mike. Flipping S2 while the amp is on can blast a jolt through the phones.

Power up, center boost/cut pot R5; set Q pot full CW (minimum Q). Don headphones, adjust listening level. Crank in EQ if needed to boost signals of interest or notch out noise. R7 varies boost/cut ±20 dB; R6 varies Q from ~1 (full CW) to ~6 (full CCW); R8 varies center frequency over the range ~440 to 4,800 Hz. The builder can shift this range during assembly by installing caps other than 0.0033 μF.

PPM Stuffing and Wiring Diagram

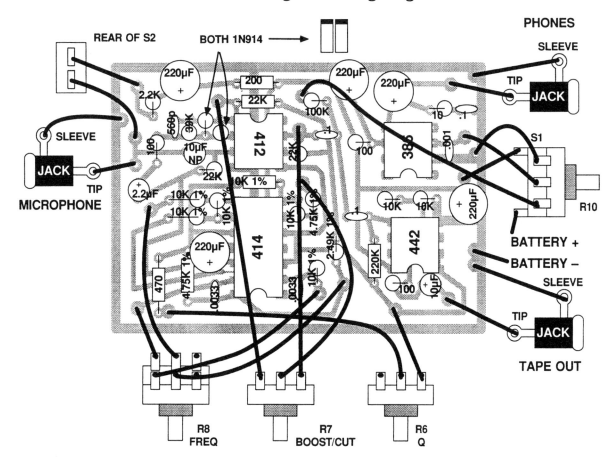

PPM Prototype

As to what tools the PPM can service, the sky is the limit: electret mikes (open, spike, hose, contact, corner); dynamic mikes, including speakers used as microphones; ceramic contact mikes; even the PZP tweeter.

BASIC "FONE" TAP

Just like the movies: access the pair, clip on wires leading to black box, roll tape.

In real life the wire specialist must prepare that black box—build himself a basic "fone" tap (BFT).

CIRCUIT FUNCTION

Line signal couples through R1-C1 and R2-C2 to inputs of U1-a, configured by R3-R4 as a differential amp with unity gain. C3 limits U1-a high-frequency response. D1 and D2 clip U1-a differential input at 1.2 V_{p-p}. Their purpose is to protect the U1-a input stage from the ring jolt. R1 and R2 show the line a high impedance and reduce the current of the ring jolt if it occurs. Note that C1 and C2 *must* be rated at least 200 volts AC.

U1-a output couples through C4 and R5 to U1-b, configured as an inverting amp with gain variable 0 to 21 by R6. C5 limits high frequency response. D3 and D4 limit U1-b output to 1.2 V_{p-p}. U1-b output couples through C6 to divider R7-R10, coupled to noninverting input of U2. C8 shunts RF at U2 input; R10 and C9 form the standard snubber. Audio couples through C10 to headphone jack.

U1-b output also couples through C13 to U1-d, configured by R13 as an AC voltage follower. Audio couples through R14 and C14 to tape output.

U1-c is configured as a DC voltage follower whose input is biased at ½ V+ by R11-R12. Its output provides a stable bias reference for U1-a/b/d.

C7, C11, C12, and R9 decouple the supply.

BFT PARTS LIST

Capacitors
C1, 2 0.1 μF coupling, 200VAC
C3 220 pF ceramic
C4, 6, 13 0.1 μF coupling
C5 100 pF ceramic
C7, 10, 11, 12 220 μF aluminum
    electrolytic
C8 0.001 μF ceramic bypass
C14 10 μF aluminum electrolytic

Resistors
R1, 2, 3, 4 22 K
R5 4.7 K
R6 100 K audio-taper pot w/switch
R7 39 K
R9 100
R10 10
R13 100 K
R14 200

Semiconductors
D1, 2 1N4007 rectifier diodes
D3, 4 1N914 or 1N4148 silicon diodes
U1 LF444 quad low-power op amp
U2 LM386 audio power driver

Miscellaneous
S1 SPST switch (part of R6)
9 V battery, wire, jacks, alligator clips,
    printed circuit board, etc.

73

BFT Schematic

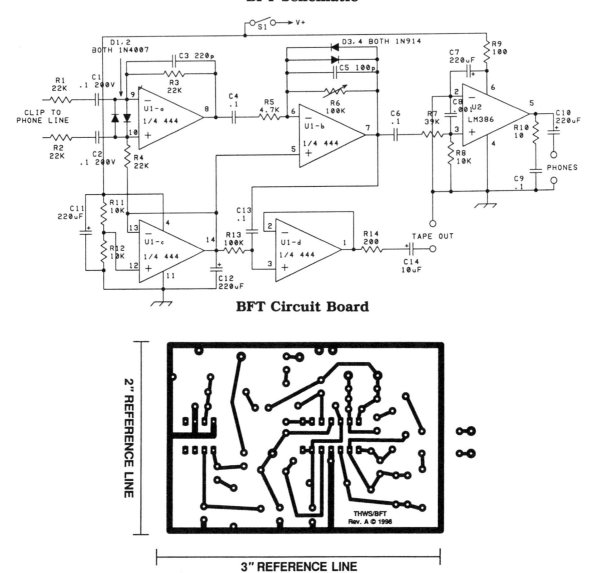

BFT Circuit Board

2" REFERENCE LINE

THWS/BFT
Rev. A © 1996

3" REFERENCE LINE

DETAILS

Construction and operation are straightforward. The BFT's high input impedance keeps it from loading the phone line but leaves it susceptible to hum from, say, a nearby fluorescent fixture. A metal enclosure and short input leads eliminate this as a practical drawback.

The BFT preamp will give several times the gain it needs to work with phone-line audio. This lets it accommodate certain types of bugs that send low-level audio out over the line while the phone is on the hook.

The BFT is not meant for long-term attachment, but for quick use against targets of opportunity. Once the pair has been accessed, the BFT clips directly to it. For greater versatility the builder can install a jack in the BFT enclosure and prepare himself a pair of plugs, one ending in loose wires with alligator clips, the other in a telco plug for that chance

BFT Stuffing and Wiring Diagram

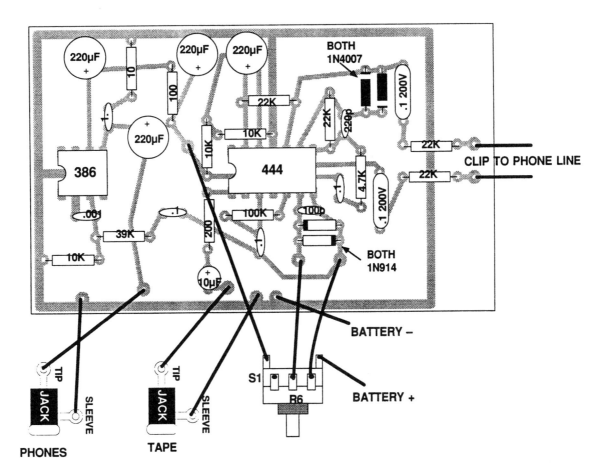

unattended outlet.

Although C1 and C2 isolate the operator from up to 200 V, they cannot guard against all hazards, such as a lightning strike or other irregular surge. User assumes *all* risk.

BFT Prototype

ANALOG AUDIO PROCESSOR

Due to quirks endemic to the world of wire specialty, raw surveillance tracks often lack the poise, grace, and general refinement needed for presentation to the grand jury. Every young specialist has had to sit, mortified, as jurors struggled to make sense of a raw feed. Given a proper intercept to start, one pass through the analog audio processor (AAP) can make it ship-shape by:

- notching out 60 Hz hum
- bandpass filtering
- notching out speech-band noise or accentuating speech-band signals with powerful parametric EQ
- taming dynamic range with variable, "studio-grade" compression

And the AAP is easy to build, straightforward to use.

CIRCUIT FUNCTION

Line-level signal couples though R1 to input level pot R2; C1 serves as an RF shunt. Signal off R2 couples through C2 to U1-a, configured by R3 and R4 as an inverting buffer with gain of ~4. C3 limits the buffer's high-frequency response; optional zener diodes D1 and D2 limit buffer output to ~11 Vp-p.

U1-a output couples directly to U1-b and -c, configured by R5-R11, C5, and C6 as a tunable notch filter designed for 60 Hz. R7 tunes frequency, R9 peaks notch depth. C5 and C6 should be the tightest tolerance

the builder can get; 5 percent types will work but should be matched with a meter if available. The notch filter is adapted from a

PARTS LIST

Capacitors
C1 220 pF ceramic
C2, 17 10 μF nonpolar electrolytic
C3, 20 100 pF ceramic
C4, 13, 15, 18 1 μF tantalum
C5, 6 0.1 μF poly, 5 percent or better (see text)
C7, 11, 12 0.01 μF poly, 10 percent or better
C8, 9 0.001 μF poly, 10 percent or better
C10 0.1 μF poly, 10 percent or better
C14, 19 10 μF aluminum electrolytic
C16 4.7 μF aluminum electrolytic
C17 0.0022 μF ceramic bypass
C21, 22 330 μF aluminum electrolytic
C23 0.1 μF ceramic bypass
(Equalizer parts not shown on schematic: 1 ea. 220 pF ceramic; 2 ea. 0.0033 μF poly, tightest tolerance builder can get; 2 ea. 1 μF tantalum caps; 1 ea. 10 μF nonpolar electrolytic)

Resistors
R1, 8, 12, 17, 40 2.2 K
R2 100 K audio-taper pot (Digi-Key p/n CT2241)
R3, 13, 14, 18, 19 22 K

R3, 13, 14, 18, 19 22 K
R4, 6 100 K
R5 47 K
R7 10 K multiturn trimpot
R9 20 K multiturn trimpot
R10, 11, 22, 30, 42 10 K
R15, 16, 20, 21, 32 39 K
R23 2.7 K
R24, 27 10 K audio-taper pot (Digi-Key
   p/n CT2238)
R25 10 K linear-taper pot (Digi-Key p/n
   CT2205)
R26 100 K dual linear-taper pot
R28, 34, 36 200
R29 3.9 K
R39, 43 36 K
(Equalizer parts not shown on
   schematic: 2   ea. 2.2 K; 6 ea. 10 K
   1 percent; 2 ea. 4.7 K; 2 ea. 100 K; 1
   ea. 2.5 K; 1 ea 470)

Semiconductors

D1, 2 5.1 V zener diode (optional)
D3, 4 1N914 or 1N4148 silicon diode
U1 TL074 quad op amp
U2, 6 TL072 dual op amp
U5 SSM2120 dynamic controller
U7 LM7805 5 V positive regulator (TO-
   220)
U8 NMH0515S 5 V to ±15 V DC con-
verter (Mouser)
(Equalizer parts not shown on
   schematic: U3 TL072 dual op amp;
   U4 TL074 quad op amp)

Miscellaneous

J1-J12 jumper
S1, 2, 3, 4 DPDT alternate-action push-
   button switch (Digi-Key p/n
EG1000)
   T1 120 V AC to 9-18 V DC wall-
   mounted, UL-approved step-down
   transformer,   500 to 1,000 ma
Heatsink for TO-220 case
RCA board-mount jacks; shielded,
   twisted-pair cable; power input jack,
   case, wire, solder, circuit board, etc.

circuit published in *Wireless World* (May 1973, p. 253).

U1-c output couples to U1-d, configured by R12-16 and C7-9 as a quasi-third-order lowpass filter, cutoff ~7 kHz. U1-d output couples to U2-a, configured by R17-21 and C10-12 as a quasi-third-order highpass filter, cutoff ~700 Hz. U2-a output couples to U2-b, configured by the ratio of R23/R22 as an inverting buffer with gain of 0.27. This gives a net passband gain of ~1.

U2-b output couples to a parametric equalizer made up of U3 and U4, the same circuit used in the PPM, modified to give greater boost/cut.

Equalizer output couples to a compressor made up of U5 and U6-a. R27 adjusts threshold over the range 100 mv to 4 V_{p-p}. Compression ratio is fixed at ~11:1.

U6-a output couples to input of U6-b, configured as a voltage follower. U6-b output couples through R28 and C14 to line output.

Each function segment—notch filter, bandpass filter, equalizer, compressor—is independently switchable IN/OUT by S1-4.

The AAP derives power from U8, a monolithic converter that steps 5 V up to ±15 V. U8 can run off any DC source that will supply 5 V (4.8 to 5.2 V) at 500 ma, here derived from U7, a 5 V regulator fed by 12 V DC (not AC) from a wall-mounted step-down transformer. U7 will accept 9 to 18 V DC.

C4, C13, C15, C18, C21, C22, and C23 decouple the supply.

DETAILS

The circuit board is laid out for board-mounting of switches and pots specified by catalog number in the parts list. Potentiometer pads have been placed relative to switch pads such that each pot shaft should protrude through a hole barely larger than its diameter. This arrangement eliminates mounting hardware but requires such accurate hole

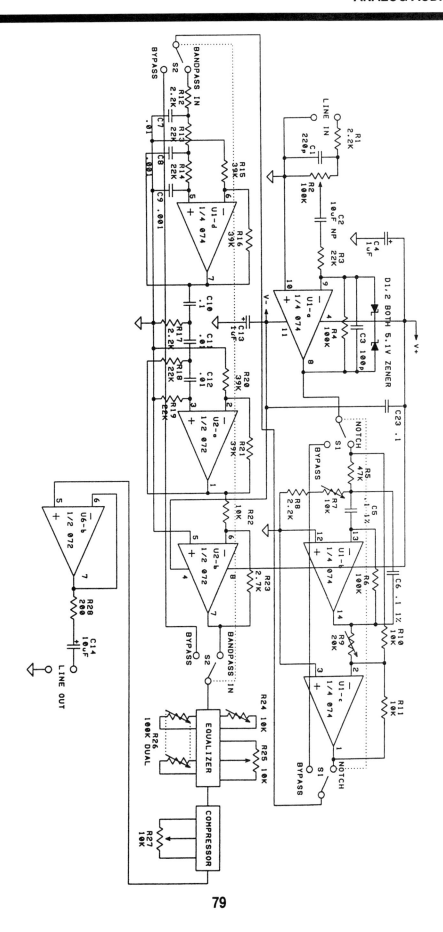

Operate/Bypass Switch Detail

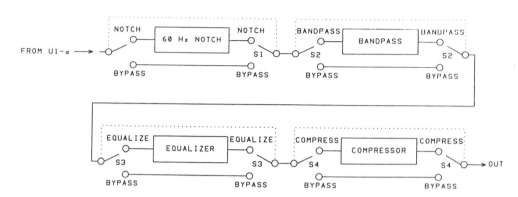

AAP Compressor Schematic

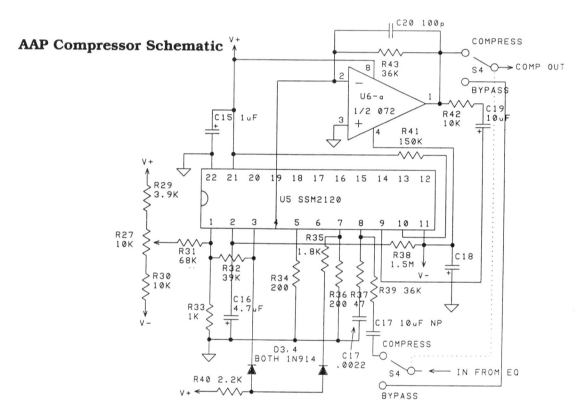

AAP Power Supply

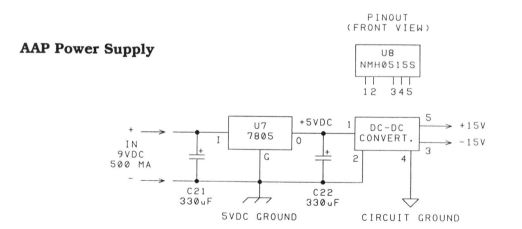

AAP Stuffing and Wiring Diagram

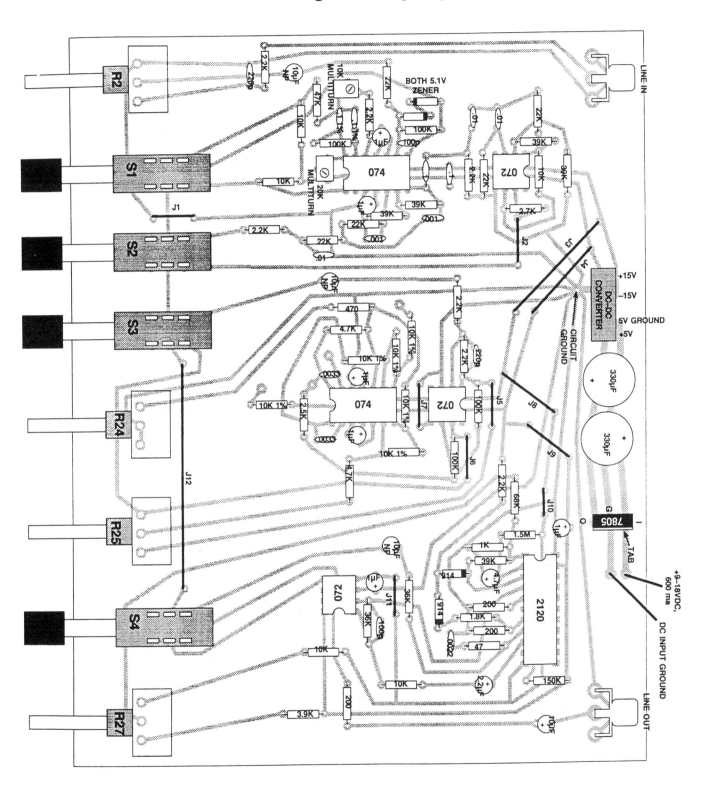

AAP Circuit Board

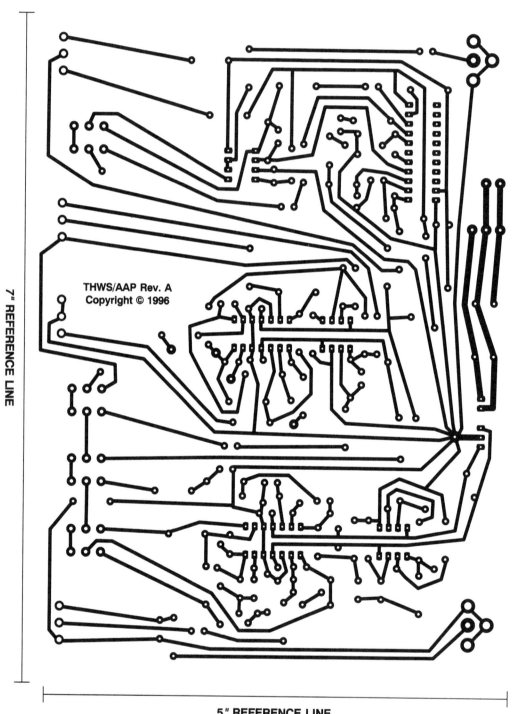

THWS/AAP Rev. A
Copyright © 1996

7" REFERENCE LINE

5" REFERENCE LINE

placement that the builder may find it advisable to make one or more cardboard test-panels, then use the best one as a drilling template.

The prototype is shown housed in a plastic case (Hosfelt p/n 14-146). Note that each shield lead on the twisted-pair cable connecting R26 to the board returns to circuit ground by a separate wire. Also, R26's metal case is independently grounded. This last measure is crucial to prevent hum when using a plastic case.

The prototype was also mounted in a metal case (Radio Shack p/n 270-274). Metal provides superior shielding for the several high-impedance nodes present in the AAP and is recommended. Metal-case mounting requires:

- that the board mount on insulated stand-offs
- that the DC input jack be insulated to isolate the 5 V DC supply from the DC converter output
- that the case connect to circuit ground at only one point
- that enough metal be cleared around line in/out jacks so that the plug can insert without touching the metal case

The 5 V regulator requires a heatsink, removed to take the prototype photo because it obscured the DC input wiring.

The prototype uses a ground plane that tests showed unnecessary if the AAP mounts in a metal case.

All chips should solder directly to the board. Wiring should follow the illustration closely.

OPERATION

After determining that the AAP is working properly, trim the notch filter. Set notch function IN, all other functions OUT. Feed a 1 Vp-p 60-Hz sinewave to the line input. Adjust R2 to give 1 Vp-p at the line output. Trim R7 to give best notch, then

Wiring Detail: R26 and Twisted-Pair Cable

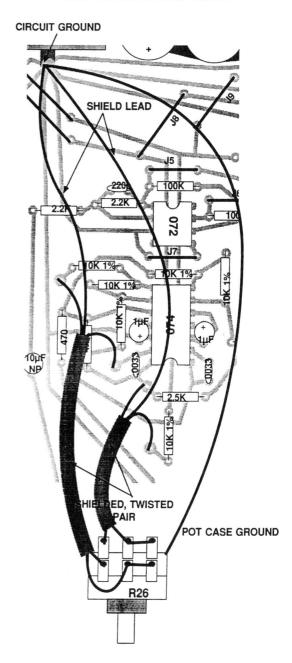

trim R9; repeat once. If no scope is available this step can be performed by ear, after feeding the AAP output into a tape deck or other audio piece equipped with a headphone driver.

The parametric equalizer gives the user independent control over frequency,

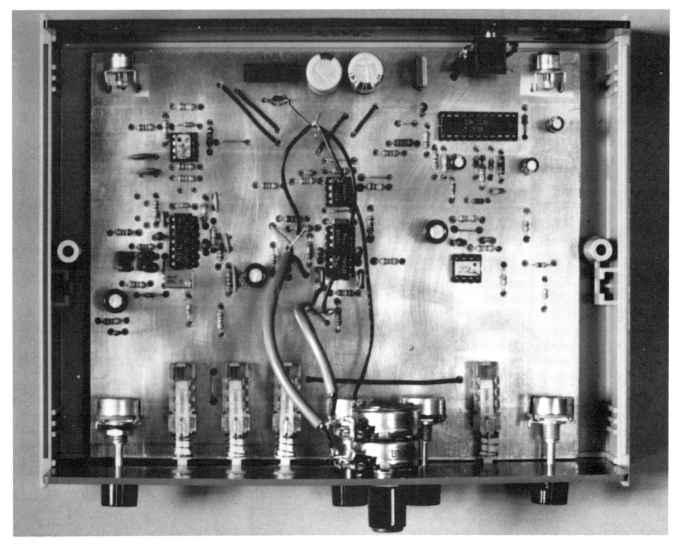

AAP Prototype

bandwidth, and boost/cut. R26 tunes over the range 450 to 4,600 Hz. R25 has no effect when centered, but will introduce up to 31 dB of boost at full CW, 31 dB of gain reduction at full CCW. The range of adjacent frequencies affected is adjustable from ~A octave (R24 full CCW) to ~1 octave (R24 full CW).

The compressor is configured for downward-only operation, ratio fixed at ~11:1, the border between heavy compression and limiting. Threshold is variable from ~4 Vp-p (R27 full CW) down to ~100 mv (R27 full CCW).

Alternate-action latching switches S1-S4

are wired such that, when the switch is physically out, its function is OUT (bypassed). Unused functions should be left OUT to minimize system noise.

Use the AAP to ameliorate several problems:

• Problem: 60-Hz hum in the source feed.
    Solution: 60-Hz notch. In severe cases harmonics of 60 Hz may also be present, in which case the bandpass filter will help, too.
• Problem: Source feed contains excessive noise above or, more typically, below the speech band.
    Solution: Bandpass filter. The builder

84

can change bandpass filter cutoff points during construction by scaling the values of C7-9 and/or C10-12.

- Problem: Excessive noise in the speech band, or too faint a speech-band signal.

    Solution: Parametric EQ. Start with boost/cut centered, bandwidth minimum. Apply boost/cut as dictated by the problem, tune R26 to zero in on the target frequency. Adjust bandwidth for optimum effect.

- Problem: Excessive dynamic range.

    Solution: Compressor. Start with threshold maximum (R27 full CW). Turn R27 CCW until signal peaks cease to exceed desired level.

The inexperienced specialist will typically test his new gadget on radio or TV sound feeds, which is unimpressive (and pointless) for demonstration purposes, because those feeds have already been compressed and equalized for maximum intelligibility. The AAP shows its stuff on raw surveillance tracks, such as tapes from the CDM or HWS or PZP.

The AAP is so simple, so powerful, that every wire specialist should have one, if only to rob himself of an excuse not to polish that raw feed before the grand jury hears it.

PARTS SOURCES

DC Electronics
Box 3203
Scottsdale, AZ 85271

LM381, NE570, MFOE71, MFOD72; 1,000
u fiberoptic cable; many other scarce chips
and discrete semiconductors.

Debco Electronics
4025 Edwards Road
Cincinnati, OH 45209

LM381, NE571, LF444; most common
semiconductors.

Digi-Key
Box 677
Thief River Falls, MN 56701

Extensive IC inventory from National
Semiconductor, Linear Technology, Maxim.
Panasonic electret condenser mikes and
ultrasonic transducers; Toko coils; 3 W
metal-oxide power resistors; power
MOSFETs (IRLD024); LN175PA and F5D1
IR LEDs; PN334PA and PN323BPA PIN
photodiodes; board-mounted pots and
switches used in the AAP; vast inventory of
general electronics.

Edmund Scientific Company
101 East Gloucester Pike
Barrington, NJ 08007

Optics, books on optics, wide variety of
general science items.

Hosfelt Electronics
2700 Sunset Blvd.
Steubenville, OH 43952

General electronics; piezo tweeter suitable
for PZP about $4; 40-kHz ultrasonic
transducers $3.50/pr. (Murata Erie
MA40A3S/R); electret condenser mikes;
NE571.

International Microelectronics
Box 170415
Arlington, TX 76003

Specializes in surplus, discontinued, and
overstocked parts. Mouser merchandise
sometimes sold at substantial discount.

Mouser Electronics
2401 Hwy. 287 N.
Mansfield, TX 76063

Miniature transformers used in HWS/HWR;
dual-ganged pots, electret condenser mikes,
extensive inventory of general electronics.

Radio Shack (found in most cities)

General electronics.

REFERENCES

Berlin, H. *Design of Phase-Locked Loop Circuits with Experiments*. Howard W. Sams, 1978.

Heyward, W. and Doug DeMaw. *Solid State Design for the Radio Amateur*. American Radio Relay League, Inc., 1986.

Jung, W. G. *Audio IC Op-Amp Applications*, 3rd Ed. Howard W. Sams, 1987.

――――. *IC Op-Amp Cookbook*, 3rd Ed. Howard W. Sams, 1986.

――――. *IC Timer Cookbook*, 2nd Ed. Howard W. Sams, 1983.

Lancaster, D. *Active-Filter Cookbook*. Howard W. Sams, 1975.

――――. *TTL Cookbook*. Howard W. Sams, 1974.

Linear Technology. *1990 Linear Applications Handbook: A Guide to Linear Circuit Design*. Linear Technology, Inc., 1989.

Maxim. *New Releases Databook, Volume III*. Maxim, Inc., 1994.

Motorola. *Linear and Interface Integrated Circuits*, Rev. 2. Motorola Semiconductor, Inc., 1988.

Signetics. *Linear Data Manual, Volume 1: Communications*. Signetics, Inc., 1987.

Sporck, C. (Ed.) *Linear Applications Handbook*. National Semiconductor Corp., 1986.

――――. *Linear Databook, Vols. 1, 2, and 3*. National Semiconductor Corp., 1986.

The Crown Boundary Microphone Application Guide. Crown International, Inc., 1990.

U.S. Patent #4,361,736. "Pressure Recording Process and Device."